ENCYCLOPAEDIA OF
ATOMIC PHYSICS

Encyclopaedia of Atomic Physics

Volume 2

Shehzad Ahmad

ANMOL PUBLICATIONS PVT. LTD.
NEW DELHI - 110 002 (INDIA)

ANMOL PUBLICATIONS PVT. LTD.
H.O.: 4374/4B, Ansari Road, Darya Ganj,
New Delhi-110 002 (India)
Ph.: 23278000, 23261597

B.O.: No. 1015, Ist Main Road, BSK IIIrd Stage
IIIrd Phase, IIIrd Block
Bangalore - 560 085 (India)
Visit us at: www.anmolpublications.com

Encyclopaedia of Atomic Physics

First Published, 2007

ISBN 81-261-3072-5 (Set)

[Responsibility for the facts stated, opinions expressed, conclusions reached and plagiarism, if any, in this book is entirely that of the Author. The Publishers bear no responsibility for them, whatsoever.]

PRINTED IN INDIA

Printed at Mehra Offset Press, Delhi.

Contents

Preface

Atom was considered as the smallest and indivisible particle of matter till some decades back. But scientists have further succeeded in dividing it too. An atom consists of three parts, electron, proton and neutron. Of them, electron is the most important one. It always remains in motion yielding energy. It is this electron, which has given birth to a new world called electronic world, which is millions of years faster than the non-electronic one.

Atomic physics relates to the physical properties and functions of atoms; their relation to other particles; effects of atoms on the mechanism of natural phenomena and vice-versa. After attempts of years, scientists succeeded in breaking the nucleus of an atom which released huge energy. It consists of two subatomic particles, called lower quark and upper quark. They are the building blocks of proton and neutron.

Now, scientists are endeavouring to break even the quark. So far, they have not succeeded but their efforts are on. They are hopeful they would succeed in their endeavour one day or the other.

Encyclopaedia of Atomic Physics is an important piece of work consisting of three volumes. It attempts at highlighting the phenomena which involve atomic physics at length. Almost every aspect has been taken note of, while exploring the vast world of atomic physics.

Hopefully, it would prove to be an asset for the scholars, students, researchers, academics and the general readers, alike.

We invite and appreciate the comments and suggestions sent by the readers.

— Editor

1

The Radioactivity

The Radiations

Radioactivity in Natural: The atomic nuclei of certain elements and their salts, such as Uranium, Thorium, Radium, Polonium, etc., emit spontaneously invisible radiation which penetrates through opaque substances, ionises gases and affects photographic plates. Such elements are said to be "radioactive", and the spontaneous emission of radiation is known as "natural radioactivity".

Rutherford studied the effect of electric and magnetic fields on the radiation emitted by different radioactive substances (Fig.). He observed that the radiation has three types of rays : one which deflect toward the negative plate, second which deflect toward the positive plate, and the third which remain undeflected in the electric field. These are called 'alpha' rays' (α–rays), 'beta rays' (β–rays) and 'gamma rays' (γ-rays) respectively. α– and β–rays are actually streams of particles; hence they are called α– and β–particles. Thus, α-particles are positively charged, α-particles are negatively charged and γ-rays are electrically neutral.

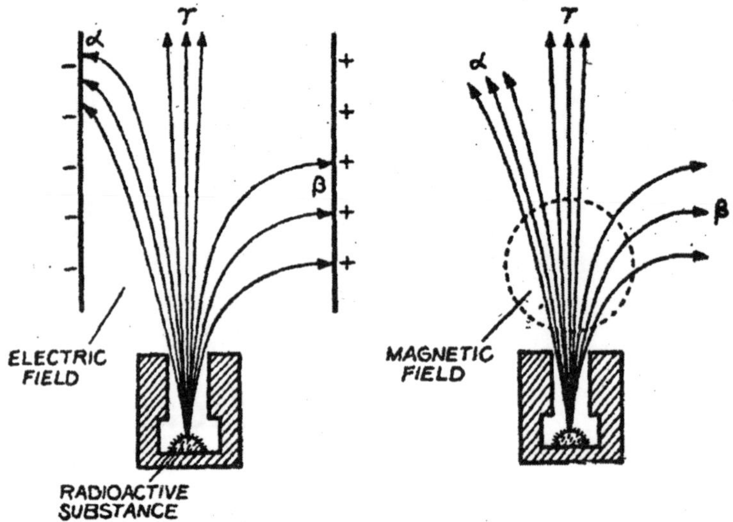

The same conclusion was drawn by passing the radiation through a magnetic field. The magnetic field in Fig. is perpendicular to the plane of paper, directed inward. In this field, α-particles are deflected to the left and β–particles to the right and move on circular paths, whereas γ-rays continue moving on their initial path. According to Fleming's left-hand rule, α-particles are positively charged and β–particles are negatively charged.

No radioactive substance emits both α- and β–particles simultaneously. Some substances emit α-particles, and some other emit β–particles. γ-rays are emitted along with α- and β–particles.

Properties of α-particles: An α-particle has a positive charge of 3.2×10^{-19} coulomb and a mass of 6.645×10^{-27} kg which are the same as the charge and mass of a helium nucleus. In fact, an α-particle is a helium nucleus and is represented by ${}_2He^4$. The main properties of α-particles are the following:

$$1000 = 2.3026\, k \log_{10} 200$$

1. α-particles are deflected in electric and magnetic fields, and the direction of deflection indicates that they are positively charged. Their small deflection (compared to β–particles) indicates that they are comparatively heavier particles.
2. The velocity of α-particles is much less than the velocity of light (less than 1/10th part), and α-particles emitted by different elements have somewhat different velocities. For example, the velocity of α-particles emitted by Uranium-1 is 1.4×10^7 meter/second and that of particles emitted by Thorium-C′ is 2.2×10^7 meter/second.
3. The range of α-particles in air (the distance travelled by an α-particle in air at NTP) varies from 2.7 cm for particles from Uranium-1 to 8.6 cm for particles from Thorium C′. In general, the range varies with the radioactive substance, and the nature and pressure of the medium.
4. α-particles can penetrate through matter but their penetrating power is small. They are stopped by only 0.1 mm thick aluminium sheet. Their penetrating power is only 1/100th of that of β–particles and 1/10,000th of that of γ-rays.
5. They cause intense ionisation in a gas through which they pass. Their ionising power is 100 times greater than that of β–rays and 10,000 times greater than that of γ-rays. Their tracks in a cloud chamber are continuous.
6. They are scattered when passing through thin foils of gold or mica. While most of the particles scatter through small angles, some of them scatter through very large angles, even greater than 90°.
7. They produce fluorescence in substances like zinc sulphide and barium platinocyanide. When a particle strikes a fluorescent screen, a scintillation (flash of light) is observed. We can count the number of α-particles by counting the number of scintillations.

8. Because of their high emitting velocity, α-particles are used for bombarding the nuclei in the transmutation of one element into other.
9. They affect a photographic plate, though feebly.
10. They produce heating when stopped.
11. They cause incurable burns on human body.

Properties of β–particles: *A* β–particle has a negative charge of 1.6×10^{-19} coulomb which is the charge of electron. Actually, β–particles are fast-moving electrons. They have the following main properties:

1. β–particles are deflected by electric and magnetic fields and the direction of deflection shows that they are negatively charged. Their deflection is much larger than the deflection of α-particles. This shows that β-particles are much lighter than the α-particles.
2. Their velocity is from 1% to 99% of the velocity of light (only in velocity, β-particles differ from cathode rays). There is enough variation in the velocities of β-particles emitted by the same radioactive substance. This is why enough dispersion is found in these particles in electric and magnetic fields (Fig.).
3. Since the velocity of β-particles is of the order of the velocity of light, their mass increases with their velocity.
4. The β-particles emitted by the same radioactive substance has a continuous distribution of kinetic energy between zero and a certain maximum value, and this maximum value is different for different substances. Hence the range of β-particles is not definite (while the range of α-particles is definite).
5. The penetrating power of β-particles is about 100 times larger than the penetrating power of α-particles. They can pass through 1 mm thick sheet of aluminium.

6. β-particles produce ionisation in gases. Their ionising power is much smaller than that of α-rays, because the mass of β-particle is much less. As β-particles cannot produce ionisation continuously, their tracks in cloud chamber do not appear to be continuous.
7. They are readily scattered while passing through matter.
8. They produce fluorescence in calcium tungstate, barium platinocyanide, etc.
9. They affect photographic plate more than the α-particles.

Properties of γ-rays: Like X-rays, the γ-rays are electromagnetic waves (or photons) of very short wavelength, ≈ 0.01 Å, which is about 1/100th part of the wavelength of X-rays. The important properties of γ-rays are as follows:

(1) γ-rays are not deflected by electric and magnetic fields. This indicates that they have no charge.

(2) They travel with the speed of light.

(3) They are most penetrating. They can pass through 30 cm thick iron sheet.

(4) They are diffracted by crystals in the same way as X-rays.

(5) They ionise gases, but their ionisation power is very small compared to that of α- and β-particles.

(6) They produce fluorescence.

(7) They affect photographic plate more than β-particles.

(8) Though there is much similarity between X-rays and γ-rays, yet their sources of origin are different. X-rays are produced by the transition of electrons in an atom from one energy level to another energy level, *i.e.,* it is an atomic property; whereas γ-rays are produced from the nucleus, *i.e.,* it is a nuclear property.

(9) These rays are absorbed by substances and give rise to the phenomenon of pair-production. When a γ-ray photon strikes the nucleus of some atom, its energy is converted

into an electron and a positron (positively-charged electron), and its own existence is extinguished:

$$\underset{(\gamma\text{-photon})}{h\nu} \rightarrow \underset{(\text{electron})}{e^-} + \underset{(\text{positron})}{e^+}$$

α-particles are doubly-charged Helium Ions: α-particles are emitted from radioactive atomic nuclei. Their deflections in magnetic and electric fields indicate that they are positively-charged. Experiments have shown that an α-particle carries a charge + 2e, which is numerically twice the charge of electron. Further, from the determination of the ratio of charge to mass, the mass of an α-particle has been calculated to be equal to the mass of a helium nucleus. From this we may conclude that α-particles may be doubly-charged helium ions, *i.e.*, helium nuclei. Rutherford and Royds, in 1909, showed experimentally that α-particles are in fact helium nuclei.

The experimental arrangement is shown in Fig. A small quantity of radon gas (radioactive substance) was sealed in a thin-walled glass tube *A*. This tube was placed in a thick-walled glass tube *B*. A capillary tube *C* provided with two electrodes was sealed on *B*. Tubes *B* and *C* were highly evacuated and the arrangement was allowed to stand for a few days.

The α-particles emitted by the radon passed through the thin walls of the tube *A* and collected in tube *B*. The gas collected in tube *B* could be forced into the capillary tube *C* by introducing mercury in *B*. After about a week enough gas was accumulated and forced into *C*.

A discharge was then passed through the gas and the emitted light was examined by a spectroscope which clearly showed spectral lines of helium. The α-particles collected in the tube *B* had caught outer electrons and had become helium atoms. This spectroscopic evidence proves conclusively that α-particles are helium nuclei.

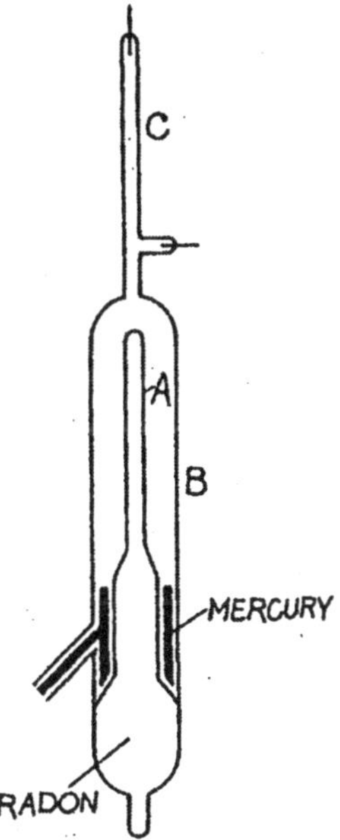

Detection of α-particles: α-particles strongly ionise air (or any other gas), and cause fluorescence in certain materials. Any of these properties may be utilised for the detection and counting of α-particles.

Two important α-particle detectors are spark counter (based on ionisation) and scintillation counter.

Range of α-particles: The distance through which an α-particle travels in a substance before coming to rest is called the 'range' of the particle in that substance. The α-particles from any one isotope are all emitted with approximately the same energy and have a well-defined range which is characteristic of that isotope.

When an α-particle passes through matter say air, it gradually loses its energy chiefly in inising the air molecules. At the start of its path, the particle may produce 20,000-30,000 ion pairs per cm of air. (This number can be found by measuring the density of the particle track obtained in a cloud chamber). The ionisation increases as the particle loses speed. This is because the slowing-down particle spends more and more time in the vicinity of the molecules. Finally, the unionisation reaches a maximum, when as much as 70,000 ion pairs per cm are produced, and then falls sharply as the particle comes to a stop (Fig.). The total distance travelled by the α-particle before coming to rest is the range R of the particle. Most α-particle sources have ranges between 2.7 and 8.0 cm in air.

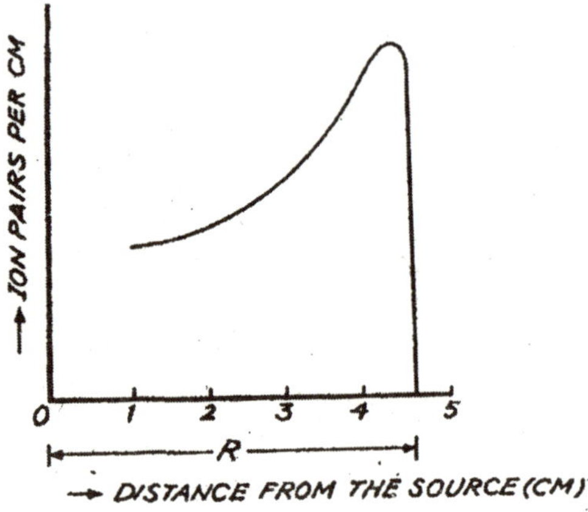

Relation of Range and Energy

The range of an α-particle in a substance is related to its initial kinetic energy. The energy lost by the particle per unit path in the substance is called the 'stopping power' $S(E)$ of the substance:

$$S(E) = -\frac{dE}{dx}$$

The stopping power varies with the energy of the particle, and the range of the particle is given by

$$R = \int_{E_0}^{0} \frac{dE}{S(E)} = \int_{E_0}^{0} \frac{dE}{-dE/dx},$$

where E_0 is the initial kinetic energy.

The stopping power $S(E)$ of a substance can be determined by measuring (e.g., by magnetic deflection) the energy of the particles after they have travelled a certain distance in the substance. When the energy loss is deter, mined for different initial velocities, the range in the substance can be deduced from the above relation as a function of the initial energy. A graph between range in air and initial energy is shown in Fig.

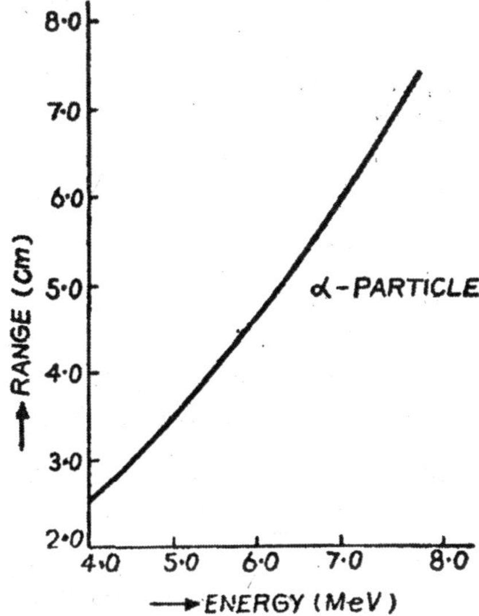

Experimental Determination of Range of α-particles: Experimentally, the range of α-particles in air can be measured using either a spark gap or a vertical ionisation chamber of adjustable height (Fig.). A radioactive source of α-particles is placed on a small platform inside the chamber whose top can be adjusted. A suitable constant potential difference is applied across the chamber, and the resulting ionisation current is measured by means of an electrometer.

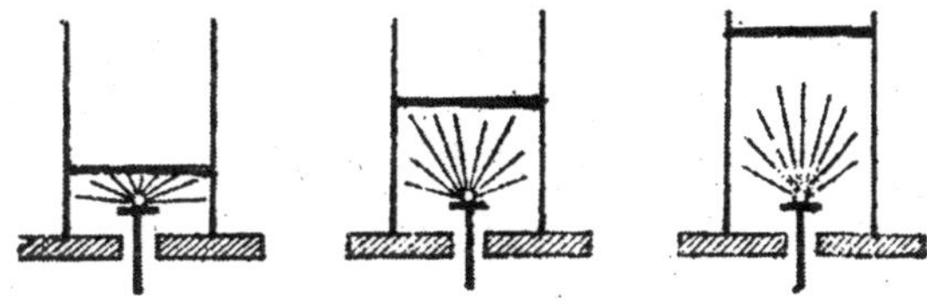

As the top of the chamber is raised, the current at first increases with the height of the chamber and then becomes constant (Fig.). In the beginning when the height is very small (Fig.), the tracks of the α-particles inside the chamber are short so that only few pairs of ions are produced.

As the height of the chamber is increased (Fig.) the tracks are lengthened so that more pairs of ions are formed and the current increases. However, when the height is greater than a certain value (Fig.), no further pairs of ions are produced because the range of the α-particles in air is limited and the lengths of the tracks cannot be increased beyond this limited range. The distance between the source and the top when the curve levels off gives an approximate value for the range of these α-particles in air. In Fig. the mean range is *OR*.

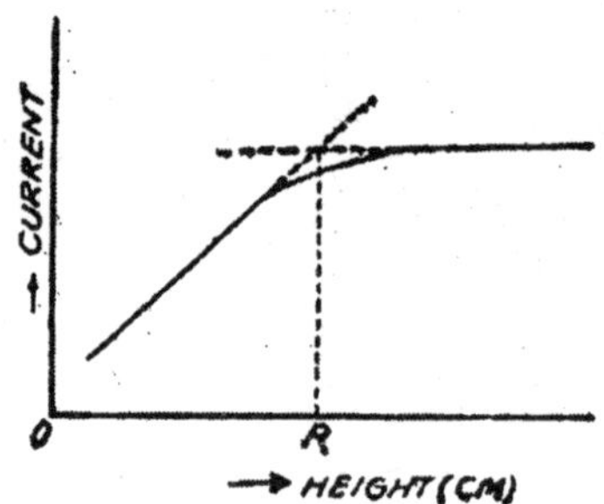

All the α-particles from a given radioactive source have the same range and hence the same energy. A close study, however, shows that α-particles often fall into a few close but sharp energy groups. This is known as the 'fine structure' of the α-rays. It gives the clue that there are different energy levels within the nucleus, *i.e.* the nucleus may exist in one or more excited states above the ground state. The existence of these excited states also explain the origin of λ rays from a radioactive nucleus.

Theory of Geiger and Nuttall

Different α-emitters emit α-particles of different energies, and hence of different ranges. It has been found that the

α-emitters giving the higher-energy particles have the shortest half-lives (or largest decay constants), and vice-versa.

Geiger and Nuttall obtained an empirical relation between the decay constant λ of an α-emitter and the range R (in air) of the α-particles emitted by it. This relation which applies to members of a particular radioactive series is

$$\log \lambda = a + b \log R,$$

where a and b are constants for the given radioactive series. This law can be derived by quantum mechanics.

A few short-lived α-emitters, thorium C' and radium C', emit α-particles of unusually long range (10-12 cm) occasionally. Such a long-range α-particle is emitted from the parent nucleus when it is in an excited state, so that it carries not only its normal energy but also the excitation energy of the parent nucleus.

Radioactive Reactions

Radioactive substances emit spontaneously either α-particles or β-particles, and some times γ-rays also. It is due to the disintegration of the nuclei of the atoms of the radioactive substance. This spontaneous emission is called 'radioactive disintegration' or 'radioactive decay'.

Rutherford-Soddy Theory of Radioactive Disintegration: Rutherford and Soddy studied the radioactive decay and formulated a theory which is based on the following laws:

(i) The radioactive emission is characteristic of the isotope, it varies from one isotope to another of the same element.

(ii) The emission occurs spontaneously and cannot be speeded up or slowed down by physical means such as change of pressure or temperature.

(iii) The disintegration occurs at random and which atom will disintegrate first, is simply a matter of chance.

(iv) The rate of disintegration of a particular substance (i.e. number of atoms disintegrating per second) at any instant is proportional to the number of atoms present at that instant.

Let N be the number of atoms present in a radioactive substance at any instant t. Let dN be the number that disintegrates in a short interval dt. Then the rate of disintegration is $-dN/dt$, and is proportional to N, *i.e.*,

$$-\frac{dN}{dt} = \lambda N.$$

where λ is a constant for the given substance and is called 'decay constant' (or 'disintegration constant' or 'radioactive constant' or 'transformation constant').

The above relation can be written as

$$\frac{dN}{F} = -\lambda \; dt.$$

Integrating, we get

$$\log_e N = -\lambda t + C,$$

where C is the integration constant. To determine C, we apply the initial conditions. Suppose there were N_0 atoms in the beginning, *i.e.* $N = N_0$ at $t = 0$. Then

$$\log_e N_0 = C$$

$$\therefore \qquad \log_e N = -\lambda t + \log_e N_0$$

$$\text{or} \qquad \log_0 \frac{N}{n_0} = -\lambda t$$

$$\text{or} \qquad \frac{N}{N_0} = e^{-\lambda t}$$

$$\text{or} \qquad \boxed{N = N_0 \, e^{-\lambda t}.}$$

This equation shows that the number of atoms of a given radioactive substance decreases exponentially with time (*i.e.* more rapidly at first and slowly afterwards). This is the Rutherford-Soddy law of radioactive decay.

The graph between the number of atoms left in a substance and the time is shown in Fig.

Statistical Nature of Radioactive Decay: Measurements of the rate of decay of a radioactive substance are purely statistical averages based on measurements made with a large number of atoms. Every atom in a sample of radioactive substance has a certain probability of decaying, but we cannot know which atom will actually decay in a particular time-interval.

Furthermore, an atom has no memory of the past. Suppose a particular atom has a probability of decay of 1 in 10^6 in a given length of time.

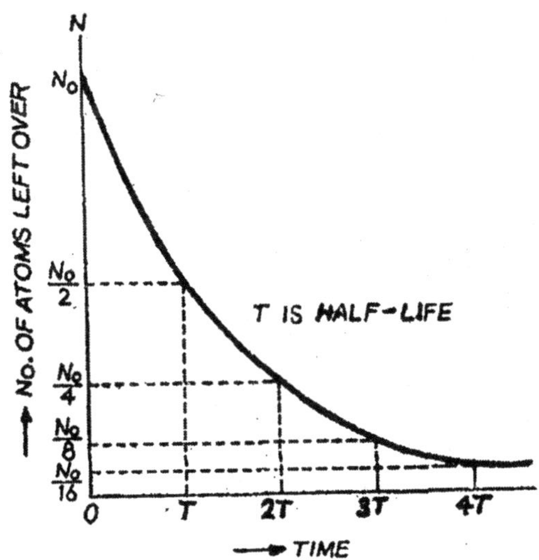

After a thousand years, if it has not decayed, it still has the same probability of decay. Thus for an atom the decay probability per unit time is constant until it actually decays.

Half-life: The atoms of a radioactive substance undergo continuous decay so that their number goes on decreasing. The time-interval T in which the mass of a radioactive substance, or the number of its atoms, is reduced to half its initial value is called the 'half-life' of that substance. The half-life of a radioactive substance is constant, but it is different for different substances. It can be read from the graph in Fig.

Relation between Half-life and Decay Constant: Let N_0 be the number of atoms present in a radioactive substance at time $t = 0$, and N the number at a later time t. Then, by Rutherford-Soddy law, we have

$$N = N_0 e^{-\lambda^i},$$

where λ is the decay constant for the substance.

Now, at $t = T$ (half-life period), $N = \frac{1}{2} N_0$. Therefore,

$$\frac{N_0}{2} = N_0 e^{-\lambda T}$$

or $$\frac{1}{2} = e^{-\lambda T}$$

or $$e^{\lambda T} = 2$$

or $$\lambda T = \log_e 2$$

or $$\boxed{T = \frac{\log_e 2}{\lambda} = \frac{0.693}{\lambda}.}$$

This is the relation between half-life and decay constant.

The half-life of a radioactive substance cannot be changed by any physical or chemical means. The half-life of an isotope of Lead ($_{82}Pb^{214}$) is 26.8 minutes. If this isotope forms some compound by chemical combination, even then its half-life will be 26.8 minutes.

The half-life of natural radioactive substances and their isotopes vary from a fraction of a second to hundreds of millions of years. The half-life of an isotope of Polonium ($_{84}Po^{214}$) is 10^{-5} second, whereas the half-life of Uranium ($_{92}U^{238}$) is 4.5×10^9 years.

The determination of half-lives is very useful for geologists for estimating the ages of mineral deposits, rocks and earth. We can also calculate how long our present stock of radioactive substances will last.

Average Life (or Mean Life) of Radioactive Atom: The atoms of a radioactive substance are continuously disintegrating. Which atom will disintegrate first is a matter of chance. Atoms which disintegrate in the beginning have a very short life and those which disintegrate at the end have the longest life. The average life of a radioactive atom is equal to the sum of the life-times of all the atoms divided by the total number of atoms.

Relation between Mean-life Time and Decay Constant: Suppose N_0 is the total number of atoms at $t = 0$, and N the number remaining at the instant t. Suppose a number dN of them disintegrates between t and $t + dt$. As the interval dt is small, we may assume that each of these dN atoms had a life-time of t seconds. Thus the total life-time of dN atoms is equal to $t\, dN$.

Since the disintegration process is a statistical one, any single atom may have a life from 0 to ∞. Hence the sum of the life-times of all the N_0 atoms is

$$\int_{t=0}^{t=\infty} t\, dN.$$

Dividing it by N_0, the total number of atoms originally present, we get the average life-time $\overline{T}$ of an atom.

Therefore

$$t = \infty$$

$$\overline{T} = \frac{1}{N_0} \int_{t=0}^{t=\infty} t\, dN.$$

Now, from the disintegration law, we have

$$N = N_0 e^{-\lambda t}$$

or

$$dN = -N_0 \lambda e^{-\lambda t} dt$$

Substituting in the above expression, we get

$$\overline{T} = \frac{1}{N} \int_{t=0}^{\infty} t\left(N_0 \;{}^{-\lambda t}\; dt\right) = \lambda \int_{0}^{\infty} t\, e^{-\lambda t}\, dt.$$

Integrating by parts, we get

$$\overline{T} = 1\left[\frac{t\, e^{-\lambda t}}{-\lambda}\right]_0^\infty - \lambda \int_0^\infty \frac{e^{-\lambda t}}{-\lambda} dt$$

$$= \left[-te^{-\lambda t}\right]_0^\infty + \int_0^\infty e^{-\lambda t}\, dt$$

The first term, on substituting the limits becomes zero. Therefore

$$\overline{T} = \int_0^\infty e^{-\lambda t}\, dt = \left[\frac{e^{-\lambda t}}{-\lambda}\right]_0^\infty$$

$$= \left[0 - \frac{1}{-\lambda}\right] \qquad [\therefore\ e{-}\infty = 0]$$

$$= \frac{1}{\lambda}.$$

Thus, the mean-life $\overline{T}$ of a radioactive atom is equal to the reciprocal of its disintegration constant.

Relation between Mean-life and Half-life: The mean-life time $\overline{T}$ of a radioactive atom is different from the half-life T. We have seen that

$$\overline{T} = \frac{1}{\lambda}$$

and $$T = \frac{0.693}{\lambda}$$

$$\therefore \quad \overline{T} = \frac{T}{0.693} = 1.44\ T$$

Thus mean life is longer than half-life. The reason is that the last few atoms of a radioactive substance may last for a very long period of time.

Series of Radioactivity

Practically all natural radioactive elements lie in the range of atomic numbers from Z=83 to Z=92. The nuclei of these elements are unstable and disintegrate by ejecting either an α-particle or a β-particle. The ejection of an α-particle lowers the mass number *A* by 4, and atomic number Z by 2. The ejection of a β-particle has no effect on mass number but increases the atomic number by 1. The atomic number is characteristic of an element and a change in it implies the formation of an atom of a new element. The new atom so formed is also radioactive and further disintegrates into another new atom, and so on. Thus a series of new radioactive elements is produced by successive disintegrations until a stable element is obtained. Such a series is called a 'radioactive series'.

There are four important radioactive series: (*i*) Uranium series, (*ii*) Thorium series, (*iii*) Actinium series and (*iv*) Neptunium series.

Uranium Series: In this series the parent element is Uranium with atomic number Z = 92 and mass number A = 238 ($_{92}U^{238}$). The sequence starts by the loss of one α-particle giving the product ${}_{90}Th^{234}$, an isotope of thorium. This is followed by the emission of two β-particles which brings its nuclear charge to the original value 92, producing an isotope of Uranium, ${}_{92}U^{234}$. This, in turn, emits an α-particle thus producing another isotope of Thorium ${}_{90}Th^{230}$ which then emits another α-particle and becomes Radium ${}_{88}Ra^{226}$.

The end-product of this series, after the emission of five more α-particles and four more β-particles, is radium-lead (${}_{82}Pb^{206}$), which is indistinguishable chemically from ordinary lead. It is a stable isotope of lead.

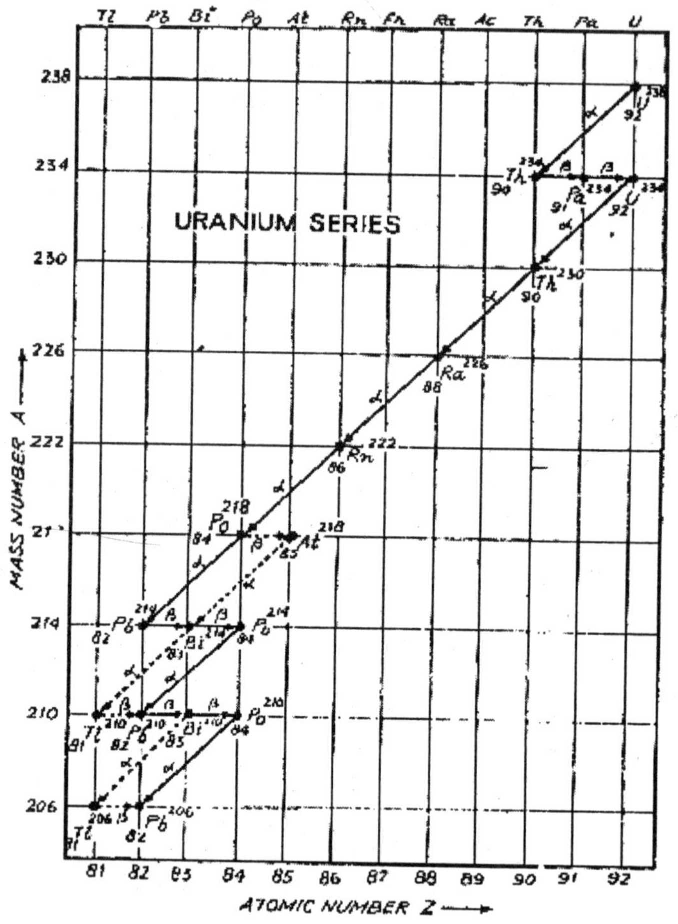

Several isotopes occurring in the series, such as ${}_{84}Po^{218}$, ${}_{83}Bi^{214}$, and ${}_{83}Bi^{210}$, have been found to decay in an alternative way also, as shown by dotted arrows in Fig.

Thorium Series: The parent element of this series is Thorium with Z = 90 and *A* = 232 (${}_{90}Th^{232}$). It goes through a series of transformations in many respects similar to the Uranium series, and end with an stable isotope of lead (${}_{82}Pb^{208}$).

Actinium Series: The parent element in this series is an isotope of Uranium called Actino-Uranium (${}_{92}U^{235}$). The end product is again an stable isotope of lead (${}_{82}Pb^{207}$).

Neptunium Series: With the discovery of the unstable transuranium element, another radioactive series was traced out.

This is called the 'Neptunium series' after its longest-lived member Neptunium. Its origin can be traced back to Plutonium. It does not end in a stable isotope of lead, but in the stable isotope of Bismuth (${}_{83}Bi^{209}$).

Radioactive Displacement Laws

(i) When a radioactive atom emits an α-particle, the product atom shifts in the periodic table two steps in the direction of lower atomic number and its mass number is lowered by 4 units.

(ii) On the emission of a β-particle, the product atom shifts one step in the direction of increasing atomic number.

These are called the 'rules of radioactive displacement'.

Radioactive Series Growth and Decay (Successive Radioactive Disintegrations): When a pure sample of radioactive atoms is isolated, it does not remain pure. The parent atom decays into a daughter atom, which decays in turn, and so on, until finally a stable end-product is reached. If the parent atom has a long half-life, the daughters, grand-daughters, and great-grand-daughters are all present with it. We can determine their numbers that exist in the mixture at a specific time.

Let N_0 be the number of parent atoms isolated at time $t = 0$. Let these atoms be denoted by 1. Suppose these atoms are decaying into atoms of a second kind, denoted by 2, which in turn, are decaying into atoms of a third kind, denoted by 3. Let the atoms 3 be stable end-products. Suppose the numbers of atoms of the three kinds 1, 2 and 3 at anytime t are N_1, N_2 and N_3 respectively, and the disintegration constants are λ_1, λ_2 and λ_3. According to the basic law, the rate of decrease of the parent atoms 1 (by their decay into atoms 2) at time t is given by

$$-\frac{dN_1}{dt} = \lambda_1 N_1. \qquad \text{... } (i)$$

The rate of increase of the atoms 2 is equal to their rate of production $\lambda_1 N_1$ from the decay of parent atoms minus the rate of their own decay $\lambda_2 N_2$ into the atoms 3. Thus

$$\frac{dN_2}{dt} = \lambda_1 N_1 - \lambda_2 N_2. \qquad (ii)$$

The rate of increase of the (stable) atoms 3 is equal to the rate of decay of the atoms 2 into the atoms 3. Thus

$$\frac{dN_3}{dt} = \lambda_2 N_2. \qquad (iii)$$

From Rutherford-Soddy equation, the number N_1 of parent atoms 1 at the time t can be directly written as

$$N_1 = N_0 e^{-\lambda_1 t}. \qquad \text{... } (iv)$$

Substituting this value of N_1 in eq. (*ii*), we get

$$\frac{dN_2}{dt} = \lambda_1 N_0 e^{-\lambda_1 t} - \lambda_2 N_2$$

or

$$\frac{dN_2}{dt} + \lambda_2 N_2 = \lambda_1 N_0 e^{-\lambda_1 t}$$

Multiplying both sides by $e^{-\lambda_2 t}$, we obtain

$$e^{\lambda_2 t} \frac{dN_2}{dt} + \lambda_2 N_2 e^{\lambda_2 t} = \lambda_1 N_0 e^{(\lambda_2 - \lambda_1)t}$$

or $$\frac{d}{dt}\left(N_2 e^{\lambda_2 t}\right) = \lambda_1 N_0 e^{(\lambda_2 - \lambda_1)t}$$

Integrating: $$\lambda_2 e^{\lambda_2 t} = \frac{\lambda_1}{\lambda_2 - \lambda_1} N_o e^{(\lambda_2 - \lambda_1)t} + C,$$

where C is a constant of integration.

At $t = 0$, $N_2 = 0$ (because only parent atoms 1 were present initially).

$$\therefore \qquad C = -\frac{\lambda_1}{\lambda_2 - \lambda_1} N_o.$$

This gives

$$N_2 e^{\lambda_2 t} = \frac{\lambda_1}{\lambda_2 - \lambda_1} N_o e^{(\lambda_2 - \lambda_1)t} - \frac{\lambda_1}{\lambda_2 - \lambda_1} N_0.$$

or $$\mathrm{N}_2 = \frac{\lambda_1}{\lambda_2 - \lambda_1} \mathrm{N}_0 \left(e^{-\lambda_1 t} - e^{-\lambda_2 t}\right). \qquad \text{... } (v)$$

On substituting this value of N_2 in eq. (*iii*) and then integrating and applying the condition that at $t = 0$, $N_3 = 0$ we get

$$N_3 = N_0\left(1 + \frac{\lambda_1}{\lambda_2 - \lambda_1} e^{-\lambda_2 t} - \frac{\lambda_2}{\lambda_2 - \lambda_1} e^{-\lambda_1 t}\right). \qquad \text{... } (vi)$$

Eq. (*iv*), (*v*) and (*vi*) represent the numbers of atoms 1, 2 and 3 respectively present in the mixture at a specified time t.

Fig. shows the numbers of atoms 1, 2 and 3 as a function of time when it is assumed that atoms 2 have a half-life longer than atoms 1, while atoms 3 are stable. Initially there are N_0 parent atoms. The number N_1 of the parent atoms l decreases exponentially according to eq. (*iv*). The number N_2 is initially zero, it increases, passes through a maximum, and then decreases gradually in accordance to eq. (*v*). The number of atoms N_3 of the stable end-product increases steadily with time, eventually approaching N_0.

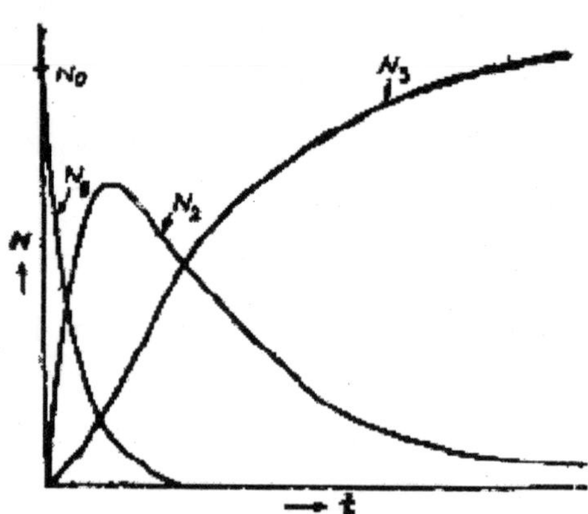

Equilibrium of Transient Radioactive

In a, radioactive series in which the parent has a very long half-life, a state is reached after a fairly long time when each daughter-product is formed at the same rate as it decays. When this condition is reached, the proportions of the different radioactive atoms in the mixture do not change with time, and the parent is said to be in 'secular radioactive equilibrium' with its daughter-products. We can prove it from the theory of successive disintegrations:

Suppose that the parent atoms 1 have a very long half-life (*i.e.* decay much more slowly) than any of the decay products ($T_1 >> T_2$). Then

$$\lambda_1 \approx 0, \text{ and } \lambda_1 << \lambda_2, \lambda_1 << \lambda_3, \text{ and so on.}$$

Substituting $\lambda_1 \approx 0$ in eq. (iv), we get

$$N_1 \approx N_0. \qquad [\therefore \; e^{-\lambda_1 t} \approx 1 \text{ when } \lambda_1 \approx 0]$$

Substituting $\lambda_1 << \lambda_2$ in eq. (v); we get

$$N_2 = \frac{\lambda_1}{\lambda_2} N_0 \left(1 - e^{-\lambda_2 t}\right) \quad \left[\because \frac{\lambda_1}{\lambda_2 - \lambda_1} \approx \frac{\lambda_1}{\lambda_2}\right]$$

$$= \frac{\lambda_1}{\lambda_2} N_1 \left(1 - e^{-\lambda_2 t}\right) \qquad [\because N_1 \approx N_0]$$

After a sufficient period of time, *i.e.* when t becomes sufficiently large, $e^{-}\lambda^{2T}$ becomes negligibly small, so that

$$N_2 = \frac{\lambda_1}{\lambda_2} N_1$$

or $$\lambda_1 N_1 = \lambda_2 N_2.$$

$\lambda_1 N_1$ is the rate of production of the atoms 2 due to the decay of the parent atoms *l*, and $\lambda_2 N_2$ is their own rate of decay into the atoms 3, and these two rates have become equal.

Thus after a sufficient time, the activities of parent and daughter become equal. The condition is known as 'secular equilibrium' and is shown in Fig.

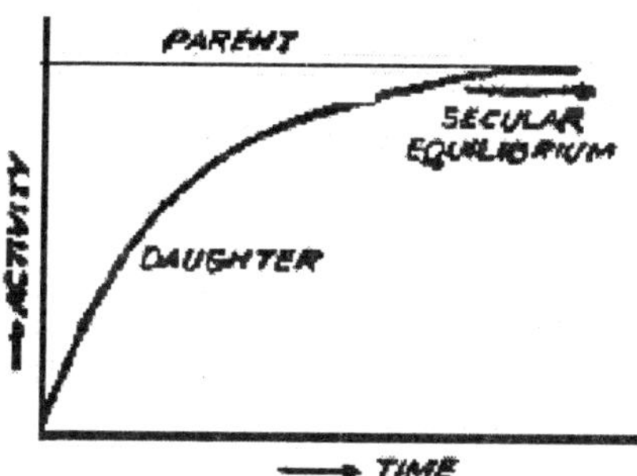

For a series of several daughter-products the above equation can be expanded as

$$\lambda_1 N_1 = \lambda_2 N_2 = \lambda_3 N_3 = \dots \dots$$

or in term of half-lives (T = log, $2/\lambda$);

$$\frac{N_1}{T_1} = \frac{N_2}{T_2} = \frac{N_3}{T_3} = \dots \dots$$

The secular equilibrium has been established in the uranium series, the (parent) uranium having a half-life (4.5×10^9 years) so long that it takes 50×10^6 years for its quantity in a rock to change by 1%.

A different type of equilibrium exists when the parent is longer-lived than the daughter ($\lambda_1 < \lambda_2$), but the half-life of the parent is not very long ($T_1 \simeq T_2$). In this case, λ_1 cannot be assumed zero. After t becomes sufficiently large, $e^{-\lambda_2 t}$ becomes negligible compared with $e^{-\lambda_1 t}$ so that the number of the daughter atoms is given by (see eq. v)

$$N_2 = \frac{\lambda_1}{\lambda_2 - \lambda_1} N_0 e^{-\lambda_1 t}$$

Comparing it with eq. (iv) and remembering that $\lambda_1 < \lambda_2$, we find that the daughter eventually decays with the same half-life as the parent. Since $N_0 \; e^{-\lambda_1 t} = N_1$, we have

$$N_2 = \frac{\lambda_1}{\lambda_2 - \lambda_1} N_1$$

or
$$\frac{N_1}{N_2} = \frac{\lambda_2 - \lambda_1}{\lambda_1}.$$

Thus after a sufficient time the ratio of parent atoms to daughter atoms becomes constant, and both eventually decay. This condition is called 'transient equilibrium', and is shown in Fig.

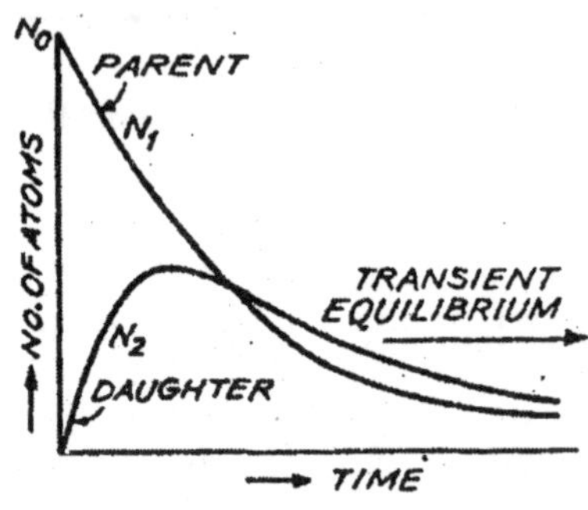

When the parent has a shorter half-life time than the daughter ($T_1 < T_2$ or $\lambda_1 > \lambda_2$), no state of equilibrium is attained. If initially we have only the parent atoms, then as the parent atoms decay, the daughter atoms increase in number, pass through a maximum, and eventually decay with their own half-life.

Activity of Radioactive Substances: The activity of a sample of any radioactive material is the rate at which its constituent atoms disintegrate. Thus if dN be the number of atoms which disintegrate during a time-interval dt, the activity R of the sample is given by

$$R = -\frac{dN}{dt}$$

The negative sign indicates that the number of atoms is decreasing with time.

If at any time the number of atoms in the sample is N, then it is an experimental fact that the rate of disintegration of the atoms is proportional to N, that is,

$$-\frac{dN}{dt} \propto N = \lambda N,$$

where λ is the decay constant for the particular atom (rather the particular isotope). Thus

$$R = \lambda N.$$

Thus the activity of the sample depends upon the number of atoms in the sample, *i.e.* upon the mass of the sample and upon the type of atom.

The traditional unit of activity is the 'curie' (*Ci*). It is defined as:

$$1 \text{ curie } (Ci) = 3.7 \times 10^{10} \text{ disintegrations/sec.}$$

1 curie is approximately the activity of one gram of radium. The smaller units are 'millicurie' (mCi) and 'microcurie' (μCi).

$$1\ mCi = 10^{-3}\ Ci = 3.7 \times 10^{7}\ \text{disintegrations/sec.}$$

$$1\ \mu Ci = 10^{-6}\ Ci = 3.7 \times 10^{4}\ \text{disintegrations/sec.}$$

Another unit of activity is 'rutherford', which is by definition

$$1\ \text{rutherford} = 10^{6}\ \text{disintegrations/sec.}$$

The SI unit of activity is the 'becquerel' (Bq); named after the discoverer of radioactivity:

$$1\ \text{becquerel (Bq)} = 1\ \text{disintegration/sec. so that}$$

$$1\ Ci = 3.7 \times 10^{10}\ \text{Bq.}$$

***Determination of Half-life* (*T*):** The half-lives of the known radioactive substances cover an enormous range from about more than 10^{15} years to about less than 10^{-6} second. Although many of these can be determined by direct measurements in the laboratory; those which are extremely long or short are estimated by less direct methods.

The moderate half-lives can be determined by measuring the activity of the given sample for a reasonable time-interval. We know that the activity of a sample at an instant t is given by

$$R = -\frac{dN}{dt} = \lambda N,$$

where N is the number of radioactive atoms in the sample at time t, and λ is the disintegration constant. From Rutherford-Soddy law,

$$N = N_0 e^{-\lambda t},$$

where N_0 is the number of atoms in the sample at time $t = 0$.

$$\therefore \quad R = \lambda N_0 e^{-\lambda t},$$

But λN_0 is the initial activity R_0 at time $t = 0$.

Therefore

$$R = R_0 e^{-\lambda t}.$$

On taking logarithms:

$$\log_c R = \log_e R_0 - \lambda t$$

or $$2.3026 \log_{10} R = 2.3026 \log_{10} R_0 - \lambda t$$

or $$\log_{10} R = \log_{10} R_0 - 0.434\ \lambda t.$$

Hence, when the logarithm of the measured activity R is plotted against the time t, a straight line is obtained whose slope is equal to $-0.434\ \lambda$. Thus, by measuring the slope, the disintegration constant λ and hence the half-life $T\ (= 0.693/\lambda)$ can be calculated.

The activity (number of atoms disintegrating per second) is measured by a Geiger counter or a scintillation counter which counts the particles emitted per second from the radioactive sample, and each emitted particle represents the disintegration of one atom.

For determining very long lives, the activity R in a given number of radioactive atoms N is measured in specially-designed counting systems and the following relation is used:

$$R = \lambda N$$

so that $$\lambda = \frac{R}{N}.$$

N is determined by finding the total mass M of the radioactive substance $\left(N = \frac{M}{\text{mass number}} \times \text{avogadro's number}\right)$.

The half-life of a very short-lived radioactive atom can be obtained, if it emits α-particles, by measuring the range of these particles and applying Geiger-Nuttall rule.

Radioactive Dating (Age of Earth): The decay of radioactive elements and its complete independence from physical and chemical conditions gives us a method for estimating the ages of old rocks and earth's crust. If a radioactive element and its decay products remain trapped in a rock (since it first solidified), the measurement of the proportion of the end-product to the parent element enables us to determine the age of the rock.

It is believed that the element U^{288} was formed at some time in the past when the earth was solidified and has been remained trapped in the rocks and earth's crust. Since then it has been decaying forming finally a stable isotope of lead (Pb^{206}), and has reached a state of secular equilibrium. Now, the disintegration constant of U^{288} is 1.54×10^{-10} $year^{-1}$. It means that, 1 gm of U^{238} produces an amount of Pb^{206} equal to

$$\frac{206}{238} \times (1.54 \times 10^{-10}) = 1.33 \times 10^{-10} \text{ gm/year.}$$

Hence, if (Pb^{206}/U^{238}) be the mass of Pb^{206} per gram of U^{233} in a specimen of earth's crust, then the time T that has elapsed since the disintegration of U^{238} started is given by (neglecting the decrease in the amount of U^{238})

$$T = \frac{\left(Pb^{206} / U^{238}\right)}{1.33 \times 10^{-10}} = (Pb^{206}/U^{288}) \times (7.5 \times 10^{9}) \text{ years.}$$

This can be taken to be the age of the earth. Measurements on the oldest rocks has established the age of the earth to be $\simeq 4 \times 10^{9}$ years.

PROBLEMS

1. *The isotope $_{92}U^{238}$ successively undergoes eight alpha decays and six beta decays. What is the resulting isotope ?*

Solution: The isotope $_{92}U^{238}$ has mass number 238 and atomic number 92. An α-decay reduces the mass number by 4 and atomic number by 2. A β-decay simply increases the atomic number by 1. Hence the new mass number is 238—(8 × 4) = 206, while the new atomic number is 92 – (8 × 2) + (6 × 1) = 82. The new isotope is $_{82}Pb^{206}$.

2. ***The thorium decay series begins with $_{90}Th^{232}$ and ends with $_{82}Pb^{208}$. How many α- and how many β-decays occur in the series ?***

Solution: An α-decay reduces the mass number by 4 and atomic number by 2. A β-decay simply increases the atomic number by 1. In the series $_{90}\text{Th}^{232} \rightarrow {}_{82}\text{Pb}^{208}$, the mass number is reduced by (232 - 208) = 24. Hence there occur 6 α-decays.

6 α decays reduce the atomic number by 12. But, in the series, the atomic number is reduced by only (90 – 82) = 8, *i.e.*, there is also an increase of 4 in the atomic number. Hence there occur 4 β-decays.

3. ***Calculate the radioactive constant for an element whose half-life period is 20 years.***

Solution: The radioactive constant λ and the half-life period T of a radioactive element are related as

$$\lambda = \frac{\log_e 2}{T} = \frac{0.693}{T}$$

Here T = 20 years.

$$\therefore \qquad \lambda = \frac{0.693}{20 \text{ years}} = 0.03465 \text{ per year.}$$

4. ***Find the half-life time and mean-life time of a radioactive substance of which the decay constant is 4.28×10^{-4} per year.***

Solution: The half life is

$$T = \frac{0.693}{\lambda} = \frac{0.693}{4.28 \times 10^{-4} \text{ / year}}$$

$$= 1619 \text{ years.}$$

The mean-life is

$$\bar{T} = \frac{1}{\lambda} = \frac{1}{4.28 \times 10^{-4} \text{ / year}} = 2336 \text{ years.}$$

5. *Ten milligrams of a radioactive substance of half-life period two years is kept in store for four years. How much of the substance remains unchanged ?*

Solution: The time-interval in which the mass (or number of atoms) of a radioactive element decays to one half of its initial value, is called the 'half-life' of the element. Thus, if the initial quantity of a radioactive substance be N_0, then the quantity N of the substance left after n half-lives is given by

$$N = N_0 \left(\frac{1}{2}\right)^n .$$

Here $\qquad N_0 = 0 \text{ mgm and } n = \frac{4 \text{ years}}{2 \text{ years}}.$

Therefore the mass of the substance remaining after 2 half-lives is

$$N = 10 \text{ mgm} \left(\frac{1}{2}\right)^2 = 2.5 \text{ mgm}.$$

6. *The half-life of a radioactive nucleus is 2.5 days. What % of the original substance will have disintegrated in 7.5 days?*

Solution: If N_0 be the initial quantity of a radioactive substance, then the quantity N left over after n half-lives is given by

Here $\qquad n = \frac{7.5 \text{ days}}{2.5 \text{ days}} = 3$

$$\therefore \qquad \frac{N}{N_0} = \left(\frac{1}{2}\right)^3 = \frac{1}{8}.$$

The fraction disintegrated is $1 - \frac{1}{8} = \frac{7}{8}$, that is, 87.5%.

7. *The half-life of radon is 3.82 days. What fraction of a freshly separated sample of this nuclide will disintegrate in*

one day ? If the sample contains initially 2.71 × 10^{15} atoms, how many atoms will disintegrate during the first day ?

Solution: If N_0 be the number of atoms at $t = 0$, and N the number at time t, then by Rutherford-Soddy law

$$N = N_0 e^{-\lambda t}$$

or $$\log_e \frac{N_0}{N} = \lambda t.$$

Here $$\lambda = \frac{0.693}{T} = \frac{0.693}{3.82} \text{ per day and } t = 1 \text{ day.}$$

$$\therefore \quad 2.3026 \log_{10} \frac{N_0}{N} = \frac{0.693 \times 1}{3.82}$$

or $$\log_{10} \frac{N_0}{N} = \frac{0.693}{3.82 \times 2.3026} = 0.0788$$

or $$\frac{N_0}{N} = \text{antilog } (0.0788) = 1.199$$

or fraction disintegrated,

$$1 - \frac{N}{N_0} = 1 - \frac{1}{1.199} = \frac{0.199}{1.199}$$

$$= 0.166 = 16.6\%.$$

If $N_0 = 2.71 \times 10^{15}$ atoms, then the number of disintegrated atoms during the first day is

$$N_0 - N = N_0 \left(1 - \frac{N}{N_0}\right)$$

$$= 2.71 \times 10^{15} \times 0.166$$

$$= 4.5 \times 10^{14}.$$

8. The half-life of Na^{24} is 15 hr. How long does it take for 87.5% of this isotope to decay ?

Or

The half-life of a radioactive substance is 15 hr. Calculate the period in which 12.5% of the initial quantity of the substance will be left over.

Solution: Let t be the time period in which 87.5% of the initial quantity of the radioactive substance is decayed, or 12.5% is left over. Now,

$$\text{half-life, } T = 15 hr,$$

$$\therefore \text{ disintegration constant, } \lambda = \frac{0.693}{15} = 0.0462 \text{ per hr.}$$

If N_0 be the number of atoms at $t = 0$, then the number N at time t is given by (Rutherford-Soddy law)

$$N = N_0 e^{-\lambda t}$$

Here $$\frac{N}{N_0} = 12.5\%$$

$$= \frac{12.5}{100} = \frac{1}{8}.$$

$$\therefore \quad \frac{1}{8} = e^{-\lambda t}$$

$$e^{\lambda t} = 8.$$

or $$\lambda t = \log_e 8 = 2.3026 \log_{10} 8$$

$$\therefore \quad t = \frac{2.3026 \log_{10} 8}{\lambda}$$

$$= \frac{2.3026 \times 0.9031}{0.0462} = 45 \text{ hr.}$$

9. The half-life of $_{11}Na^{24}$ is 15 hours. How long will it take for 93.75% of a sample of this isotope to decay ?

[Ans. 60 hr.**]**

10. ***The half-life of radon is 3.8 days. After how many days will only one-twentieth (1/20) of a radon sample be left over?***

Hint: $\frac{N}{N_0} = \frac{1}{20}$. **[Ans.** 16.5 days.]

11. ***The half-life of radium is 1590 years. In how many years will 1 gm of radium (i) be reduced to 1 centigram, (ii) loses 1 centigram.***

Solution: (*i*) Let t be the time in which 1 gm of radium will be reduced to 1 centigram ($\frac{1}{100}$ =0.01 gm). The decay constant is

$$\lambda = \frac{0.693}{T} = \frac{0.693}{1590} \text{ per year.}$$

If N_0 be the number of atoms at $t = 0$, and N the number at time t then, by Rutherford-Soddy law

$$N = N_0 e^{-\lambda t}$$

Here $$\frac{N}{N_0} = \frac{1 \text{ centigram}}{1 \text{gm}} = 0.01 = \frac{1}{100}.$$

$$\therefore \quad \frac{1}{100} = e^{-\lambda t}$$

or $$e^{\lambda t} = 100$$

or $$\lambda t = \log_e 100 = 2.3026 \times \log_{10} 100$$

$$= 2.3026 \times 2$$

$$\therefore \quad t = \frac{2.3026 \times 2}{\lambda}$$

$$= \frac{2.3026 \times 2}{0.693/1590} = 10566 \text{ years.}$$

(*ii*) In this case in the beginning (at t=0) the radium is 1 gm. In a time t (say), it loses 1 centigram (= 0.01 gm), that is, it remains (1–0.01) = 0.99 gm.

Thus $\frac{N}{N_0} = \frac{0.99}{1} = \frac{99}{100}.$

$$\therefore \quad \frac{99}{100} = e^{-\lambda t}$$

$$\text{or} \quad e^{\lambda t} = \frac{100}{99}$$

$$\text{or} \quad \lambda t = \log_e \frac{100}{99}$$

$$= 2.3026 \log_{10} \frac{100}{99}$$

$$= 2.3026 \times 0.0044.$$

$$\therefore \quad t = \frac{2.3026 \times 0.0044}{\lambda}$$

$$= \frac{2.3026 \times 0.0044}{0.693/1590} = 23.2 \text{ years.}$$

12. ***10 gram of a radioactive substance is reduced by 2.5 mg in 6 years through α-decay. Evaluate half-life time and mean-life time of the substance.***

Solution: 10 gm of the substance is reduced to (10 – 0.0025) = 9.9975 gm in 6 years.

Let N_0 be the amount of the substance at $t = 0$ and N at time t. Then, we have

$$\frac{N}{N_0} = e^{-\lambda t},$$

where λ is decay constant. Here, $\frac{N}{N_0} = \frac{9.9975}{10}$ and $t = 6$ years.

$$\therefore \quad \frac{9.9975}{10} = e^{-6\lambda}$$

$$\text{or} \quad e^{6\lambda} = \frac{10}{9.9975}$$

$$\text{or } \lambda = \frac{1}{6} \times 2.3026 \times (\log_{10} 10\text{-}\log_{10} 9.9975)$$

$$= \frac{1}{6} \times 2.3026 \times (1.0000 - 0.9999)$$

$$= 3.84 \times 10^{-5} \text{ per yr.}$$

$$\therefore \text{ half-life, } T = \frac{0.693}{\lambda} = \frac{0.693}{3.84 \times 10^{-5}} = 1.8 \times 10 \text{ years}$$

$$\text{and mean-life, } \overline{T} = \frac{1}{\lambda} = \frac{1}{3.84 \times 10^{-5}} = 2.6 \times 10^4 \text{ years.}$$

13. ***A radioactive element disintegrates for an interval of time equal to its mean life. (i) What fraction of element remains? (ii) What fraction has disintegrated ?***

Hint: $\overline{T} = 1/\lambda$.

[**Ans.** (*i*) $1/e$, (*ii*) $(e - 1)/e$]

14. ***$_{83}Bi^{212}$ decays to $_{81}Tl^{208}$ by α-emission in 34% of the disintegrations and to $_{84}Po^{212}$ by β-emission in 66% of the disintegrations. If the total half-value period is 60 5 minutes, find the decay constants for α and β emissions.***

Solution: Certain nuclei break up in two different ways, either by α-emission or by β-emission, giving rise to two different product nuclei. The probability of disintegration is the sum of separate probabilities and

$$\lambda = \lambda_\alpha + \lambda_\beta.$$

$$\text{The half-life is } T = \frac{0.693}{\lambda} = \frac{0.693}{\lambda_\alpha + \lambda_\beta}.$$

Thus, here

$$\lambda_\alpha + \lambda_\beta = \frac{0.693}{T} = \frac{0.693}{60.5 \text{ min}} = 0.01145 \text{ min}^{-1}.$$

Since $\frac{dN}{dt} = \lambda N$, we can write

$$0.34 = \lambda_\alpha N$$

and $$0.66 = \lambda_\beta N.$$

These give $$\frac{0.34}{0.66} = \frac{\lambda_\alpha}{0.01145 - \lambda_\alpha} . [\because \lambda_\alpha + \lambda_\beta = 0.01145]$$

Solving $$\lambda_\alpha = 0.003893 \text{ min}^{-1}$$

and $$\lambda_\beta = 0.007557 \text{ min.}^{-1}$$

15. ***The mean lives of a radioactive substance are 1620 and 405 years for α-emission and β-emission respectively. Find out the time during which three-fourth of a sample will decay if it is decaying both by α-emission and β-emission simultaneously.***

Solution: The decay constant for α- and β-emission are $\frac{1}{1620}$ and $\frac{1}{405}$ per year respectively.

$\therefore$ total decay constant, $\lambda = \frac{1}{1620} + \frac{1}{405} = \frac{1}{324}$ per year.

If N_0 be the amount of a sample at $t = 0$ and N the amount left over at time t, then by the decay law

$$N = N_0 e^{-\lambda t}$$

After $\frac{3}{4}$th part has been disintegrated, we have $N = N_0/4$. Therefore

$$\frac{1}{4} = e^{-\lambda t}$$

or $$e^{\lambda t} = 4$$

or $$t = \frac{1}{\lambda} \log_e 4 = \frac{1}{\lambda} (2.3026 \log_{10} 4)$$

$$= 324\ (2.3026 \times 0.6020) = 449 \text{ years.}$$

16. ***The activity of a radioactive substance decreases to 1/64 of its original value in 21 years. Calculate the half-life of the substance.***

Solution: The activity (rate of disintegration) at any instant is directly proportional to the number of radioactive atoms present at that instant. Thus a decrease in activity means a corresponding decrease in the number of atoms. If N_0 be the number of atoms initially and N the number left after a time-interval t (21 years), then

$$\frac{N}{N_0} = \frac{1}{64}.$$

By Rutherford-Soddy law, we have

$$\frac{N}{N^0} = e^{-\lambda t} = \frac{1}{64}$$

or $$e^{\lambda t} = 64$$

or $$\lambda t = \log_e 64$$

$$= 2.3026 \times \log_{10} 64.$$

Here $$t = 21 \text{ years.}$$

$\therefore$ $$\lambda = \frac{0.3026 \times 1.8062}{21}$$

$$= 0.198 \text{ per year.}$$

Hence the half-life is

$$T = \frac{0.693}{\lambda} = \frac{0.693}{0.198} = 3.5 \text{ years}$$

17. ***A radioactive substance at a given instant emits 4750 particles per minute. Five minutes later it emits 2700 particles per minute. Find the decay constant and half-life of the substance.***

Solution: The rate of disintegration of a substance at any instant is proportional to the number of atoms at that instant.

Thus, if N_0 be the number of atoms at $t = 0$, when the rate of disintegration is 4750 min^{-1}, and N be the number of atoms at $t = 5$ min when the rate of disintegration is 2700 min^{-1}, then

$$\frac{N}{N_0} = \frac{2700}{4750}.$$

From Rutherford-Soddy law, $N/N_0 = e^{-\lambda t}$

$\therefore \quad \lambda t = \log_e (N/N_0) = 2.3026 \log_{10} (N_0/N)$.

Here t=5 min and $N_0/N = 4750/2700 = 1.76$.

$$\therefore \qquad \lambda = \frac{2.3026 \log_{10} 1.76}{5 \text{ min}}$$

$$= \frac{2.306 \times 0.2455}{5 \text{ min}} = 0.113 \text{ min}^{-1}.$$

The half-life T of the material is

$$T = \frac{0.693}{\lambda} = \frac{0.693}{0.113} = 6.13 \text{ min}.$$

18. ***The activity of a radioactive substance is reduced to half of its original value in 3.86 days. Find the time-interval in which its activity will become 1%. What is the average life of the atom of the substance ?***

Solution: The activity is proportional to the number of atoms. It is reduced to half in 3.86 days. It means the half-life T is 3.86 days.

$$\therefore \qquad \lambda = \frac{0.693}{3.86}$$

$$= 0.1795 \text{ per day}.$$

Let t be the time in which the number of atoms reduces to 1%. By the decay law,

$$= \frac{N}{N_0} = e^{-\lambda t}$$

Here $$\frac{N}{N_0} = 1\% = \frac{1}{100}.$$

$$\therefore \qquad \frac{1}{100} = e^{-\lambda t}$$

or $$e^{\lambda t} = 100$$

or $$t = \frac{1}{\lambda} \log_e 100 = \frac{1}{\lambda} (2.3026 \log_{10} 100).$$

$$= \frac{2.3026 \times 2}{0.1795} = 25.6 \text{ days.}$$

The average life of the atom is

$$\overline{T} = \frac{1}{\lambda} = \frac{1}{0.1795} = 5.57 \text{ days.}$$

19. ***The half-life of $_{92}U^{238}$ against α-decay is 4.5×10^9 years. Find the activity of 1 gm of $_{92}U^{238}$. (Avogadro Number = 6.02×10^{28} per gin-atom).***

Solution: 1 gm-atom of U^{238} has a mass of 238 gm and contains 6.02×10^{23} atoms (the Avogadro's number). Therefore, the number of atoms in 1 gm of U^{238} is

$$N = \frac{6.02 \times 10^{23}}{238} = 2.53 \times 10^{21}.$$

Its half-life is $T = 4.5 \times 10^9$ years. Therefore its decay constant is

$$\lambda = \frac{0.693}{T} = \frac{0.693}{4.5 \times 10^9}$$

$$= 1.54 \times 10^{-10} \text{ year}^{-1}.$$

Now, the activity R, or the rate of disintegration $-\frac{dN}{dt}$ of a substance at any instant is proportional to the number of atoms N present at that instant, *i.e.*

$$R = -\frac{dN}{dt} = \lambda N$$

$$= (1.54 \times 10^{-10}) \times (2.53 \times 10^{21})$$

$$= 3.9 \times 10^{11} \text{ year}^{-1}$$

$$= \frac{3.9 \times 10^{11}}{365 \times 24 \times 60 \times 60}$$

$$= 1.23 \times 10^{4} \text{ integrations/sec}$$

$$= 1.23 \times 10^{4} \text{ Bq.}$$

20. ***Calculate the radioactivity in curie for 1 gm of $_{38}Sr^{90}$ with half-life period of 28 years. Given: Avogadro number = 6.025 × 10^{26} molecules/kg molecule.***

Solution: The intensity of radioactivity for 1 gm of substance is given by

$$R = \lambda N,$$

where λ is decay constant and *N* is the number of atoms in 1 gm.

Here $$\lambda = \frac{0.693}{\text{half-life}}$$

$$= \frac{0.693}{28 \text{ years}} = \frac{0.693}{28 \times 365 \times 24 \times 60 \times 60 \text{ sec}}$$

$$= 7.85 \times 10^{-10} \text{ sec}^{-1}.$$

and $$N = \frac{6.025 \times 10^{23} \text{ atoms/gm-atom}}{90 \text{ gm/gm-atom}}$$

$$= 6.69 \times 10^{21} \text{ atoms/gm.}$$

∴ $$R = (7.85 \times 10^{-10} \text{ sec}^{-1})(6.69 \times 10^{21} \text{ atoms})$$

$$= 5.25 \times 10^{12} \text{ disintegrations/sec.}$$

Now, 1 curie = 3.7 × 10^{10} disintegration/sec.

∴ $$R = \frac{5.25 \times 10^{12}}{3.7 \times 10^{10}} = 142 \text{ Ci.}$$

21. ***One gm of Ra^{226} has an activity of 1 curie. Determine the half-life of radium. Avogadro number is 6.02×10^{38}.***

Solution: 1 gm-atom of radium has a mass of 226 gm, and contains 6.02×10^{23} atoms (the Avogadro's number). Therefore, the number of atoms in 1 gram of radium is

$$N = \frac{6.02 \times 10^{23}}{225} = 2.66 \times 10^{21}.$$

Now, the activity R of a radioactive sample is the rate of disintegration of its atoms;

$$R = -\frac{dN}{dt} = \lambda N,$$

where λ is the disintegration constant.

Here R = 1 curie = 3.7×10^{10} disintegrations/sec and $N = 2.66 \times 10^{21}$.

$$\therefore \qquad \lambda = \frac{R}{N} = \frac{3.7 \times 10^{10}}{2.66 \times 10^{21}} \doteq 1.39 \times 10^{-11} \text{ sec}^{-1}$$

The half-life is

$$T = \frac{0.693}{\lambda} = \frac{0.693}{1.39 \times 10^{-11}} = 4.94 \times 10^{10} \text{ sec.}$$

Now, 1 year = $365 \times 24 \times 60 \times 60 = 3.15 \times 10^{7}$ sec.

$$\therefore \qquad T = \frac{4.98 \times 10^{10}}{3.15 \times 10^{7}} = 1581 \text{ years}$$

The average-life is

$$\overline{T} = \frac{1}{\lambda} = \frac{1}{1.39 \times 10^{-n}} = 7.19 \times 10^{10} \text{ sec}$$

$$= \frac{7.19 \times 10^{10}}{3.15 \times 10^{7}} = 2282 \text{ years.}$$

22. ***A sample of U^{234} for which the decay constant is 8.78×10^{-14} per sec undergoes 3.7×10^{8} disintegrations per second. Find the mass of the sample. Avogadro number = 6.03×10^{28}.***

Solution: The activity in N atoms is

$$R = \lambda N.$$

Here $R = 3.7 \times 10^8$ disintegrations/sec and

$$\lambda = 8.78 \times 10^{-14} \text{ per sec.}$$

$$\therefore \quad N = \frac{R}{\lambda} = \frac{3.7 \times 10^8}{8.78 \times 10^{-14}} = 4.21 \times 10^{21} \text{ atoms.}$$

Thus there are 4.21×10^{21} atoms in the given sample of U^{234}.

Mass of 6.03×10^{23} atoms of U^{234} = 234 gm.

$\therefore$ mass of 4.21×10^{21} atoms of

$$U^{234} = \frac{234 \times \left(4.21 \times 10^{21}\right)}{6.03 \times 10^{23}} = 1.63 \text{ gm.}$$

23. ***The half-life of a cobalt radio-isotope is 5.3 years. What strength will a milli-curie source of the isotope have after a period of one year and 5.3 years?***

Solution: The strength, (activity) R of a radioactive source containing N atoms is given by

$$R = \lambda N,$$

where λ is decay constant.

Here, $$\lambda = \frac{0.693}{5.3 \text{ yr}} = 0.131 \text{ per yr.}$$

Let N_0 be the number of atoms at $t = 0$ when the strength is 1 milli-curie, and N the number at a time $t = 1$ year when the strength is R' (say).

Then

$$1 = \lambda N_0 \text{ and } R' = \lambda N.$$

$$\therefore \quad \frac{N}{N_0} = R'.$$

But from Rutherford-Soddy law,

$$\frac{N}{N_0} = e^{-\lambda t}$$

$$\therefore \qquad R' = e^{-\lambda t}$$

or $$\log_e\left(\frac{1}{R'}\right) = \lambda t.$$

Here $$\lambda = 0.131 \text{ per year}$$

and $$t = 1\text{yr}.$$

$$\therefore \qquad \log_e\left(\frac{1}{R'}\right) = 0.131 \times 1.$$

or $$\log_e\left(\frac{1}{R'}\right) = \frac{0.131 \times 1}{2.3026} = 0.0569$$

or $$\frac{1}{R'} = \text{antilog } 0.0569 = 1.14$$

$$\therefore \qquad R' = \frac{1}{1.14} = 0.88 \text{ milli-curie}.$$

The strength after 5.3 years (half-life) will become one-half, *i.e.*, 0.5 milli-curie.

24. ***The isotopes U^{238} and U^{235} occur in nature in the ratio 140: 1. Assuming that at the time of earth's formation they were present in equal ratio, make an estimation of the age of the earth. Th e half-lives of U^{238} and U^{235} are 4.5×10^9 years and 7.13×10^8 years respectively. ($\log_{10} 140 = 2.1461$, $\log_{10} 2 = 0.3010$)***

Solution: Let N_1 and N_2 be the number of atoms of U^{238} and U^{235}, and T_1 and T_2 their half-lives.

From the decay formula $N = N_0^{e - \lambda t}$, we have

$$\frac{N_1}{N_2} = e^{(\lambda_2 - \lambda_1)t},$$

where t is the elasped time.

From this, we have

$$t = \frac{\log_e (N_1 / N_2)}{\lambda_2 - \lambda_1}$$

$$= \frac{\log_e (140/1)}{\log_e 2\left(\frac{1}{T_2} - \frac{1}{T_1}\right)} \qquad \left[\therefore \lambda = \frac{\log_e 2}{T}\right]$$

$$= \frac{\log_{10} 140}{\log_{10} 2}\left(\frac{T_2 T_1}{T_1 - T_2}\right)$$

$$= \frac{2.1461}{0.3010} \times \frac{7.13 \times 10^8 \times 4.5 \times 10^9}{37.87 \times 10^8}$$

$$= 6\times 10^9 \text{ years.}$$

25. ***Find the half-life of uranium, given that 3.23 × 10⁻⁷ gm of radium is found per gm of uranium in old minerals. The atomic weights of uranium and radium are 238 and 226 and half-life of radium is 1600 years. (Avogadro Number is 6023 × 10²³/gm-atom).***

Solution: The half-life of uranium is very much longer than that of radium. Therefore, in very old minerals, radioactive equilibrium exists, and we have

where N_U and N_R are the number of atoms of uranium and radium at any time, and λ_U and λ_R the corresponding decay constants.

Since $\lambda \propto \frac{1}{T}$ (where T is half-life), we can write

$$\frac{N_U}{T_U} = \frac{N_R}{T_R}$$

or $$T_U = \left(\frac{N_U}{N_R}\right). \qquad (i)$$

N_U is the number of uranium atoms in *l* gm of uranium while N_R is the number of radium atoms in 3.23×10^{-7} gm of radium. Thus

$$Nu = \frac{6.023 \times 10^{23}}{238}$$

and $$N_R = \frac{6.023 \times 10^{23}}{226} \times (3.23 \times 10^{-7}).$$

$$\therefore \quad \frac{N_U}{N_R} = \frac{226}{238 \times 3.23 \times 10^{-7}} = 2.94 \times 10^6.$$

Substituting for $\frac{N_U}{N_R}$ and for T_R (= 1600 years) in eq. (*i*), we get

$$T_U = (1600 \text{ years}) \times (2.94 \times 10^6)$$
$$= 4.70 \times 10^9 \text{ years}.$$

26. ***The atomic ratio between the uranium isotopes U^{238} and U^{234} in a mineral sample is found to be 1.8×10^4. The half-life of U^{234} is 2.5×10^5 years. Find the half-life of U^{238}***

[Ans. 4.5×10^9 years**].**

N_U is the number of uranium atoms in 1 gm of uranium while N_R is the number of radium atoms in 3.23×10^{-7} gm of radium. Thus

$$N_U = \frac{6.023 \times 10^{[illegible]}}{[illegible]}$$

and $$N_R = \frac{6.0[illegible]}{[illegible]} (3.23 \times 10^{-7}).$$

$$\frac{N_U}{N_R} = [illegible]94 \times 10^{[illegible]}.$$

Substituting for $\frac{N_U}{N_R}$ and T_R (= 1600 years) in Eq. ( ),

[illegible]

$$T_U = 1600 \times [illegible] \times 10^{6}$$

$$= 4.70 \times 10^{[illegible]} \text{ years}.$$

28. [illegible] U^{238} [illegible] U^{234} [illegible] $\times 10^{[illegible]}$ [illegible]

[illegible]

(Ans. [illegible] $\times 10^{[illegible]}$ years).

2

Nuclear Radioactivity

Decay of Radioactive Nucleus

Explanation of α-emission from Radioactive Nuclei: The nuclei of heavier atoms, beyond bismuth ($_{83}Bi^{209}$), are unstable (radioactive) with respect either to α- or to β-emission. This is because these nuclei are so large that the *short-range* nuclear forces holding the nucleons together are hardly able to counterbalance the electrostatic repulsion between the large number of protons in them, α-emission occurs in such nuclei as a means of increasing their stability by reducing their size. In fact, all nuclei with $Z > 83$ and $A>209$ spontaneously decay into lighter nuclei through the emission of α-particle which is $_2He^4$ nucleus. The equation for α-decay can be written as:

$$\underset{\text{Parent nucleus}}{_ZX^A} \rightarrow \underset{\text{Daughter nucleus}}{_{Z-2}Y^{A-4}} + \underset{\text{Alpha particle}}{_2He^4}$$

Since each α-particle is a helium nucleus composed of two protons and two neutrons, their emission from nuclei fits with the proton-neutron picture of the nucleus.

An important question is that why do the radioactive nuclei emit α-particles ($_2He^4$) rather than protons ($_1H^1$) themselves. The answer lies in the *high* binding energy of the α-particle. To escape from a nucleus, a particle must have kinetic energy. Only the α-particle mass is sufficiently smaller than the sum of the masses of its constituent nucleons. Therefore, in the formation of α-particle within the nucleus sufficient energy is released which becomes available to the particle to escape.

Again, there remains the problem of *how* an α-particle can actually escape the nucleus. A nucleus is surrounded by a potential barrier, and the escaping particle must have enough energy to cross the barrier. The uranium nucleus, for α-particles, has a potential barrier of 27 MeV height so that only particles having 27 MeV or more energy would be able to escape. But α-particles emitted by uranium have an energy of only 4 MeV. Then how they get out across the barrier at all ? The explanation is quantum-mechanical and is based on the following lines:

(i) An α-particle, before emission, exists as such within the nucleus.

(ii) It is in constant motion but is kept inside the nucleus by the surrounding potential barrier.

(iii) There is a small, but finite, probability that the particle may "leak" through the barrier each time it collides with it. (This is due to the wave nature of the particle).

(iv) Once the particle leaks through the barrier, it escapes from the nucleus because of its kinetic energy and the electrostatic repulsion.

An α-particle within a nucleus collides with the surrounding barrier about 10^{21} times per second. Calculations show that in case of U^{238} nucleus the chance of escape in any single collision is only 1 out of 10^{38}. Thus an α-particle may have to try for $\frac{10^{38}}{10^{21}} = 10^{17}$ sec $= 3 \times 10^9$ years before it actually escapes. This explains the very long half-life of U^{238} which runs in billions of years (4.5 billions).

Po^{214}, in contrast to U^{238}, has a half-life of only 0.0001 sec. There are two reasons for it: (i) An α-particle in Po^{214} nucleus has a large energy. (ii) The height of the potential barrier is smaller. Therefore the chance of escape in a single collision is much larger, 1 out of 10^{17}. Hence the mean waiting time for the escape reduces to $\frac{10^{17}}{10^{21}} = 0.0001$ sec.

Explanation of β-emission from Radioactive Nuclei: Like α-decay, the β-decay and positron emission are the means by which a nucleus alters its composition (neutron/proton ratio) to achieve greater stability.

β-decay is the mission of electrons from nuclei. All properties of nuclei, however, firmly indicate that they do not contain electrons. They are composed of protons and neutrons only. Then how electrons are emitted from radioactive nuclei? Yukawa's meson theory supplies an answer to this question.

In the nucleus the protons and neutrons are held together by the continual exchange of pi-mesons between them. The electron emission takes place on the spontaneous conversion of a neutron into a proton by ejecting a π^--meson. This π^--meson decays almost instantly into an electron (e^-) and an antineutrino (θ). The reaction is

$$n \rightarrow p + \bar{\pi} \rightarrow p + e^- + \bar{\upsilon}.$$

Since in β-decay a neutron is converted into a proton, the neutron/ proton ratio decreases.

In β-decay, the mass number (number of nucleons) of the radioactive nucleus remains unchanged, but the atomic number (number of protons) increases by unity. Hence the equation for β-decay can be written as

$$_{Z}X^{A} \rightarrow {}_{Z+1}Y^{A} + {}_{-1}\beta^{o} + \overline{v}.$$

Parent nucleus — Daughter nucleus — Beta particle (electron)

Similarly, the positron emission takes place on the spontaneous conversion of a proton into a neutron by ejecting π^{+}-meson which decays almost instantly into a positron (e^{+}) and a neutrino. The reaction is

$$p \rightarrow n + \pi^{+} \rightarrow n + e^{+} + v.$$

In this process the neutron/proton ratio increases.

In positive β-decay (positron emission) the mass number remains unchanged, but the atomic number decreases by unity. Hence the equation for positron emission can be written as

$$_{Z}X^{A} \rightarrow {}_{Z-1}Y^{A} + {}_{+1}\beta^{o} + v.$$

Parent nucleus — Daughter nucleus — Positron

The significance of the emission of antineutrino in β-decay and of neutrino in positron emission is that it keeps energy, linear momentum and angular momentum all conserved.

Capturing Electron

The electron capture is a process which is competitive with positron emission. In this process, a nucleus captures one of the inner orbital electrons of the atom, with the result that a nuclear proton is converted into a neutron and a neutrino is emitted. The reaction is

$$p + e^{-} \rightarrow n + v.$$

Usually the captured electron comes from the *K*-shell, and an *X*-ray photon is emitted.

Electron capture occurs more often than positron emission in heavy nuclei because in them the electron orbits are much nearer.

Because the absorption (capture) of an electron by a nucleus is equivalent to the emission of a positron from it, the electron capture reaction ($p + e^- \rightarrow n + v$) is essentially the same as the positron emission reaction ($p \rightarrow n + e^+ + v$).

Experimental Investigation of β-ray Energy Spectra: The β-rays emitted from radioactive nuclei consist of electrons with varying high velocities (energies). Their energy variation can be studied by a β-ray spectrometer. The radioactive material (β-ray source) is placed in vacuum upon a fine wire at S (Fig.). The emitted electrons pass through fairly wide slits and are bent by means of a magnetic field B so that they are received by a Geiger counter placed in a fixed position.

The magnetic field B is directed upwards perpendicular to the plane of paper. An electron (mass m, charge e) emitted with a velocity v and entering the field B perpendicularly, describes a circular path in the field. The magnetic force on the electron, evB, supplies the centripetal force mv^2/r, where r is the radius of the path. Thus

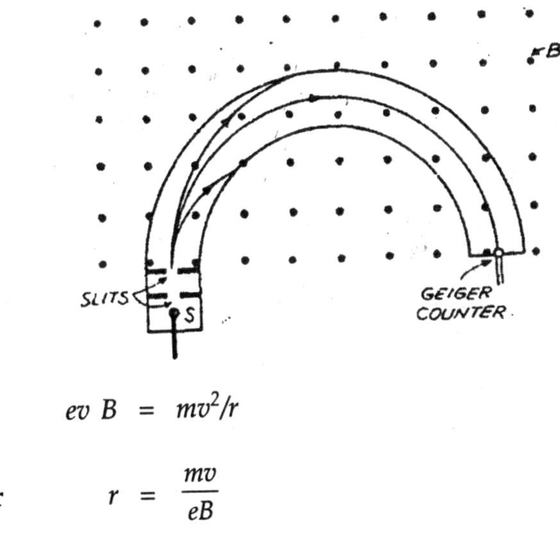

$$evB = mv^2/r$$

or

$$r = \frac{mv}{eB}$$

The position of the counter is fixed. Thus r is fixed.

Therefore, for a given value of B, only those, electrons are recorded by the counter whose momentum is given by

$$p = mv = eBr.$$

The magnetic field B is varied and the number of electrons reaching the counter per unit time is obtained for different values of B. Since each value of B corresponds to a different value of p, the numbers of electrons corresponding to different momenta are obtained. The observed momenta p can be converted into corresponding kinetic energies K from the relativist formula

$$K = V\left(m_0^2 c^4 + p^2 c^2\right) - m_0 c^2$$

where c is the speed of light. A curve between the relative number of electrons and the corresponding kinetic energy is then plotted (Fig.).

Characteristic Features of β-ray Energy Spectra: The β-particles emitted from a given radioactive nucleus have a *continuous* distribution of kinetic energies from 0 to a maximum value $K_{max.}$ The value K_{max}, called the 'end-point energy', is a characteristic of the emitter. It is 1.17 MeV for RaE (Fig.). Most of the β-particles have energies considerably less than 1.17 MeV, and the average energy per particle is about 039 MeV, only about one-third the end-point energy.

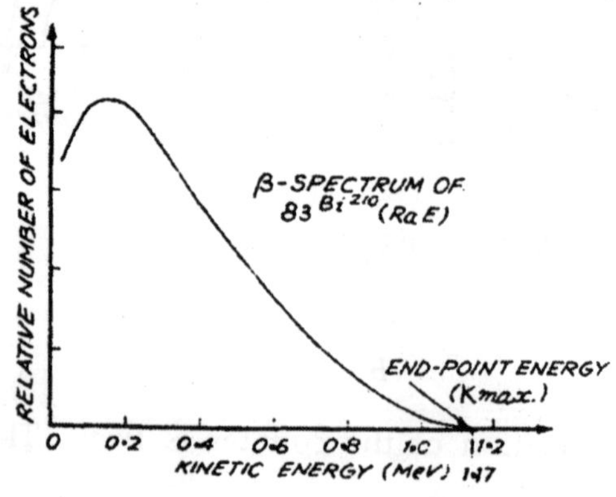

Thus the process of β-disintegration differs from α-disintegration in two important respects. Firstly, the α-particles are already present in the initial nucleus but the β-particles are *not* present in the initial nucleus and are created at the time of emission. Secondly, the energy spectrum of the β-particles is continuous, and not discrete as of α-particles.

Difficulties with β-ray Continuum: In experiments with β-decay the conservation of energy, linear momentum and angular momentum were all appeared to be violated. Let us first consider energy. A parent nucleus in a definite state of energy emits a β-particle and leaves a daughter nucleus which is also in a definite state of energy. Thus the β-particle must emit with a *fixed* energy, equal to the difference between the energies of the parent and the daughter nuclei. But, in practice, the energy spectrum of β-particles is continuous, *i.e.* a β-particle can have *any* kinetic energy between zero and K_{max}.

Meitner argued that all β-particles start from the parent nucleus with same kinetic energy K_{max} but suffer *varying* energy-loses by collision with the atomic electrons surrounding the nucleus. Hence they come out with continuously varying energies. Ellis and Wooster in 1927, performed an experiment to verify this hypothesis. They placed a β-emitting source (RaE) in a thick-walled calorimeter designed to absorb all of the emitted β-particles and measured the total heat (energy) produced by a known number of disintegrations. The heat produced divided by the number of disintegrations gave the average energy per disintegration, $K_{average}$. This was found to be 0.35 MeV, which fairly agreed with the average energy computed from the distribution curve, *but was much less than* K_{max}. Thus Meitner's hypothesis proved to be incorrect. Since the balance of energy, $K_{max} - K_{avarage}$ was unaccountable in β-disintegration, the conservation of energy appeared to be violated.

Linear and angular momenta were also found not to be

conserved in β-decay. If in a single disintegration the directions of the emitted β-particle and of the (recoiling) daughter nucleus are observed (by recording their ionisation tracks), they are not exactly opposite as required for the conservation of linear momentum.

The non-conservation of angular momentum arises because the intrinsic spins of the electron, proton and neutron are all $\frac{1}{2}$. In β-decay a neutron is converted into a proton with the emission of an electron:

$$n \rightarrow p + e^-.$$

After the decay, the proton and the electron spins can be parallel (total spin = 1) or anti-parallel (total spin = 0); but in no case the total spin can be $\frac{1}{2}$ (the spin of the original neutron). Thus the spin (and hence angular momentum) is not conserved in the above reaction.

Neutrino Hypothesis of β-Disintegration: In 1930, Pauli suggested that if a particle having zero charge, zero rest mass, and spin $\frac{1}{2}$ is supposed to be emitted together with the electron, then the energy, momentum and angular momentum would all be conserved. Fermi, in 1934, named this particle as 'neutrino' and developed a complete theory of β-disintegration. According to this theory, when a neutron (n) is converted into a proton (p); an electron (e^-) and a neutrino (v) are emitted. It was later on found that there are two kinds of neutrino, the neutrino itself (symbol $\bar{v}$) and the antineutrino (symbol $\frac{1}{2}$). In β-disintegration it is the anti-neutrino that is emitted. Thus the basic equation for β-disintegration is

$$n \rightarrow p + e^- + \bar{v}.$$

This equation removes the various difficulties encountered in β-disintegration in the following way:

(i) The neutrino (in fact antineutrino) carries no charge. This maintains conservation of charge in the β-disintegration,

(ii) The neutrino has a zero rest mass, and hence a zero rest mass energy. Therefore, in a β-disintegration, the maximum energy which can be carried off by an electron is equal to the energy equivalent of the mass-difference between the parent and the daughter nuclei. If this is E_{max}, then

$E_{max} = m_0C^2 + K_{max}$,

where m_0c^2 is the rest energy of the electron and K_{max} is the end-point (kinetic) energy of the β-ray spectrum. The energy E_{max} is distributed among the electron, the neutrino, and the recoiling daughter nucleus in a continuous range of different ways. Since the daughter nucleus carries away negligible kinetic energy, the neutrino carries off the difference between K_{max} and the actual kinetic energy of the electron for the particular disintegration. Thus the energy remains conserved in the process.

(iii) The neutrino has a momentum exactly balancing the sum of the momenta of the electron and the recoiling daughter nucleus. Thus momentum remains conserved.

(iv) Like electron, the neutrino has a spin $\frac{1}{2}$. This leads to the conservation of angular momentum because the spins of the three decay particles can combine to give $\frac{1}{2}$.

(v) Lacking mass and charge, the neutrino can pass unhindered through vast amounts of matter. If this were not so, the neutrinos would have been stopped in the calorimeter experiment and their energy would have been absorbed by the calorimeter, giving a temperature rise corresponding to K_{max}.

β-rays and Spectrum

Internal Conversion: For certain β-emitters the continuous distribution curve has a number of distinct peaks superimposed upon it (Fig.), which are the characteristics of the emitter. This means that some β-ray sources emit, in addition to the continuous spectrum, a number of β-particles of discrete energies. They are said to constitute a line spectrum of β-rays.

The line spectrum of β-rays arises due to extra nuclear electrons ejected from the atom by a process called 'internal conversion'. A nucleus in an excited state may return to a lower energy state in two ways: (i) by emitting a γ-ray photon of energy hv equal to the difference between (Fig.) the energies of the two nuclear states, (ii) by giving up the energy hv to an electron in the K-, L-......shell of the same atom. In the second case (internal conversion) the electron is ejected from the shell with a discrete kinetic energy $hv - E_k$, $hv - Ei$,......, where E_k, E_L,....... are the binding energies of, the electron in the K-, L-,....... shells respectively. (The vacancy created in the shell is filled by electrons in higher shells cascading down with the emission of X-rays characteristic of the atom.) Thus the process of internal conversion, which gives rise to β-ray line spectrum, is a *direct* transfer of excitation energy from the nucleus to one of the surrounding electrons. Since in the internal conversion the electrons do *not* come from the nucleus, the not a true form of β-disintegration.

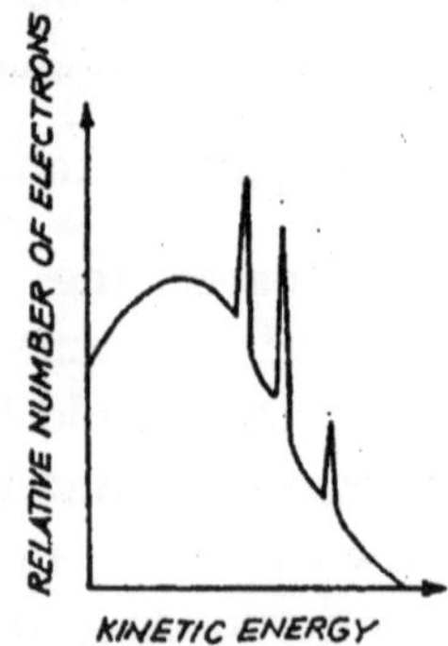

Interaction of γ-rays with Matter: The γ-rays are electromagnetic radiation of very short wavelengths (≈ 0.004 Å to 0.4 A). They have no electric charge, and so they cannot be deflected by magnetic or electric fields. Consequently, direct measurements of their energies (or wavelengths) with a magnetic spectrometer are not possible. The absorption of γ-rays by matter is also different from that of charged particles such as α- and β-rays. Charged particles lose their energy by inelastic collisions so that they slow down and finally come to rest and absorbed at the end of their range. On the other hand, when a beam of γ-ray photons passes through matter, the intensity of the beam (number of photons) decreases exponentially according to the law

$$I = I_0 e^{-\mu x}$$

where I_0 is the initial intensity of the beam, μ is the absorption coefficient of the substance and x is the thickness of the absorber. Thus γ-rays have no definite range, as do α- and β-rays.

Three separate processes are responsible for the decrease in intensity (absorption) of γ-rays. They are photoelectric absorption, Compton scattering and pair production.

(i) *Photoelectric Absorption:* In this process *all* the energy of a γ-ray photon is transferred to a bound electron, and the γ-ray photon ceases to exist. The ejected electron may either escape from the absorber or reabsorbed due to collisions. At low photon energies (5.0 keV for aluminium and 500 keV for lead), the photoelectric effect is chiefly responsible for the γ-ray absorption.

(ii) *Compton Scattering:* At energies in the neighbourhood of 1 MeV, Compton scattering becomes the chief cause of removal of photons from the γ-ray beam. In this process the γ-ray photon is scattered by one of the atomic electrons which is separated from its atom. The scattered photon moves with reduced energy in a direction different from the original direction and is thus removed from the incident beam.

(iii) *Pair Production:* At high enough energies, both the photoelectric absorption and Compton scattering become unimportant compared with pair production. In the latter process, a γ-ray photon, in passing close to an atomic nucleus in the absorbing matter, disappears and an electron and a positron are created:

$$\underset{(\gamma\text{-photon})}{\gamma} \rightarrow \underset{(\text{electron})}{e^-} + \underset{(\text{positron})}{e^+}$$

The electron and the positron form a pair of particles and hence the process is called 'pair production. The conservation of charge is obvious from the above equation. The rest mass m_0, and hence the rest mass energy m_0c^2, of the positron is the same as that of the electron, *i.e.* 0.51 MeV. The energy of the γ-ray photon, $h\nu$, must be atleast $2 \times 0.51 = 1.02$ MeV for pair production to be possible, because this amount of energy is needed to supply the rest energy of the two particles. If $h\nu$ is greater than 1.02 MeV, the balance of energy appears as kinetic energy of the particles (neglecting the small recoil energy of the nucleus).

Measurement of the Wavelengths of γ-rays: There are two chief methods for measuring γ-ray wavelengths.

(i) *By a Crystal γ-ray Spectrometer:* The method is suitable for measuring the wavelengths of the longer γ-rays. A quartz diffracting crystal *C* (Fig.) is bent and clamped so that the diffracting planes meet, when extended, in a line at *A*, normal to the plane of the figure. The radius of curvature of the crystal is then equal to the diameter of the focussing circle whose centre is *O*. Suppose the source of γ-rays is at a point *S* on the focussing circle such that the Bragg condition ($2d \sin \theta = n\lambda$) is satisfied. Then a diffracted beam enters the detector

D(a scintillation counter) as if it came from the virtual source at *S′*. For each different wavelength there is a particular position of the source on the focussing circle for which a strong diffracted beam is obtained.

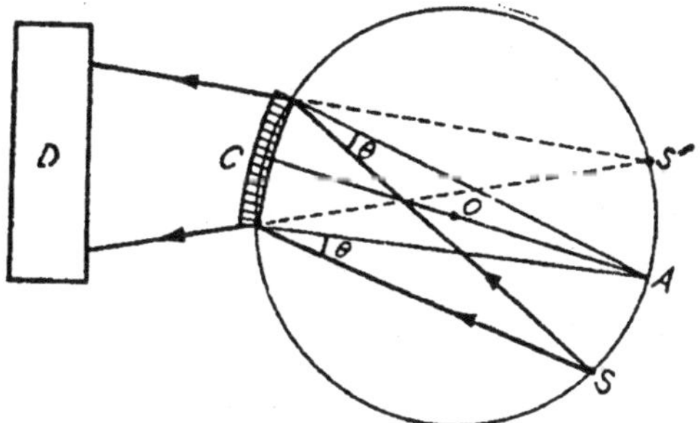

In order to perform the experiment, the source position on the circle is varied, and the corresponding counting rates of the detector are determined. A graph is plotted between the counting rate and the angle θ (position of the source). This graph is found to have sharp peaks which correspond to strong diffracted beams. The angle θ corresponding to a peak is the Bragg's angle. Thus, knowing the crystal lattice spacing d_s the wavelength A of the γ-radiation in a known order can be calculated by using Bragg's equation.

This method has two disadvantages. Firstly, the measurements become more difficult and less precise as the energy of the γ-rays increases and the wavelength decreases, and so the method can be used upto 1 MeV only. Secondly, the method requires highly active γ-ray source.

(ii) *By a Magnetic Spectrograph:* The wavelengths of γ-rays of moderate energy can be determined by an indirect method using a magnetic spectrograph. In this method we obtain a magnetic spectrum of the secondary electrons which are produced by photoelectric absorption or by Compton scattering of γ-rays in matter.

A simple arrangement, used by Ellis, is shown in Fig. The source of γ-rays, such as radium-*B*, is kept in a small thin-walled tube *S* wrapped round with a foil of lead, platinum or tungsten and placed in vacuum. The foil is thick enough to stop all the primary electrons (β-rays) from the radioactive source itself but not the γ-rays. The γ-rays passing out through the wall of the tube eject (secondary) electrons of different energies from the metallic foil due to photoelectric absorption. These electrons pass through a slit *O* and are bent by means of a magnetic field *B* perpendicular to the plane of paper so that they reach a photographic plate *P*. An electron (mass *m*, charge *e*) which comes out from the foil with a velocity *v* describes in the magnetic field a circular path of radius *r* such that the magnetic force on the electron, *evB*, supplies the necessary centripetal force mv^2/r. Thus

$$evB = mv^2/r.$$

The momentum of the electron is therefore given by

$$p = mv = eBr. \qquad \text{... } (i)$$

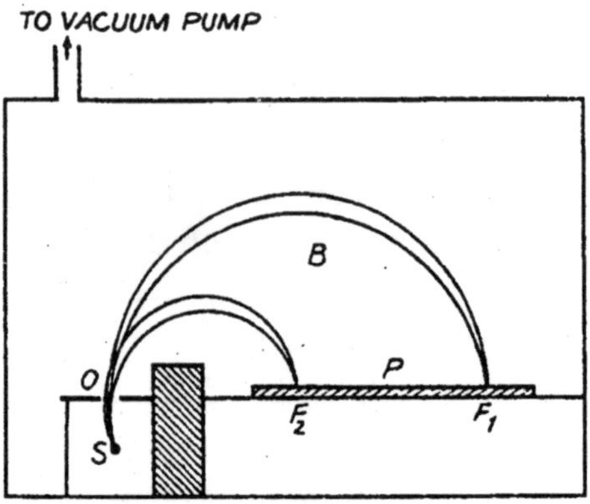

Thus those electrons which have the same momentum have circular paths of the same radii and come to a line-focus at the photographic plate. Hence upon developing the plate, a number of focal lines F_1 F_2, ... are obtained which correspond to various electron momenta (or energies).

The various momenta *p*. can be determined by knowing the magnetic field *B* and the radii *r* of the various electron paths. For example, the radius of the path corresponding to F_1 is given by

$$r = \frac{1}{2}(SF_1) = \frac{1}{2}\sqrt{(OS^2 + OF_1^2)}$$

The momenta *p* as determined from eq. (i), can Le converted into corresponding kinetic energies *K* from the relativistic formula

$$K = \sqrt{(m_0^2c^4 + p^2c^2)} - m_0c^2$$

The electrons ejected from the atoms of the metallic foil must come from one or other of the shells *K, L, M,*........ Suppose, for example, that an electron is ejected from a *K*-shell. For this, an amount of energy *Wk* is required which is supplied by the γ-ray incident upon the foil. If *hv* is the energy of the incident γ-rays, we have

$$hv = W_k + K;$$

where *K* is the kinetic energy of the ejected electron. The value of W_k for the given atom is known from X-ray studies, whereas *K* has been obtained from the magnetic spectrograph. Hence *hv,* and so v (or A), the frequency (or wavelength) of the γ-rays can be determined.

Nuclear Energy Levels—Origin of γ-rays: The discovery of the fine structure of α-rays, and also the experimental fact that γ-rays form a *line* spectrum, gave rise to the idea of nuclear energy levels. It is supposed that there are a number of discrete energy levels in the nucleus, and that a nucleus is normally in its lowest energy state, but it may also exist for short times in an excited state. The transition of the nucleus from an excited state to the lower states gives rise to γ-ray lines.

An important question which arises is how does the nucleus come in an excited state from which it can undergo transitions to lower levels ? An answer is furnished by the fact that in natural radioactivity, γ-rays are emitted only by nuclei which also emit α- or β-rays. Measurements of the α- or β-ray energies and the γ-ray energies show that the γ-rays are emitted by the *daughter* nucleus produced by the emission of an α- or β-particle from the parent nucleus.

Thus it can be supposed that the emission of an α- or β-particle sometimes leaves the daughter nucleus, not in the normal state, but in an excited state of higher energy. The excited daughter nucleus then passes to the normal state with the emission of γ-rays. If it returns to the normal state in a single transition, a single γ-photon is emitted. If it returns by a series of transitions, a series of γ-ray photons is emitted. Measurements of these γ-ray energies help in locating the energy levels of the nucleus.

As an example, radium emits α-particle to become radon in accordance with the transformation

$$_{88}Ra^{226} \rightarrow {}_{86}Rn^{222} + {}_{2}He^{4}$$

(α-particle)

The emitted α-particles fall into two *sharp* energy groups, one having an energy of 4.80 MeV and the other 4.61 MeV. Accompanying them γ-rays of 0.19 MeV energy are also detected (Fig.). The explanation is that, when radium emits an α-particle of 4.80 MeV energy, the daughter nucleus of radon is formed in its ground state. On the other hand, when radium emits an α-particle of 4.61 MeV energy, then the radon nucleus is left in an excited state, from which it passes to the ground state by emitting a γ-ray of energy 4.80 – 4.61 = 0.19 MeV. Similarly, when thorium C decays to thorium C'', five groups of α-particles are emitted, and the existence of four nuclear excited states of thorium C above the ground state is established.

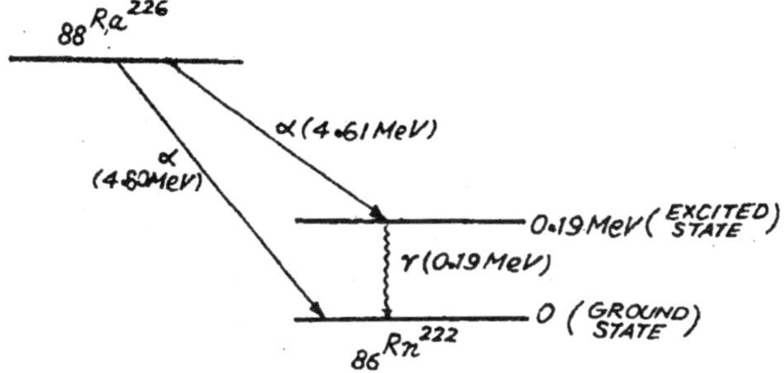

Like α-emission, β-emission also gives information about the energy levels of the daughter nuclei. For example, $_{12}Mg^{27}$ emits β-particle to become $_{13}Al^{27}$:

$$_{12}Mg^{27} \rightarrow {}_{13}Al^{27} + {}_{-1}e^{0}$$

(β-particle)

The emitted β-particles fall into two energy groups, one having an end-point (maximum) energy of 1.78 MeV and the other 1.59 MeV. (In each group the energies are distributed from zero to a maximum). Accompanying them γ-rays of energies 0.834 MeV, 1.015 MeV and 0.181 MeV are also detected. The decay scheme consistent with all of these data is shown in Fig.

The $_{12}Mg^{27}$ nucleus, after emitting the β-particle, leaves the daughter $_{13}Al^{27}$ nucleus in either of its two excited states from which it proceeds to the ground state by γ-ray emission.

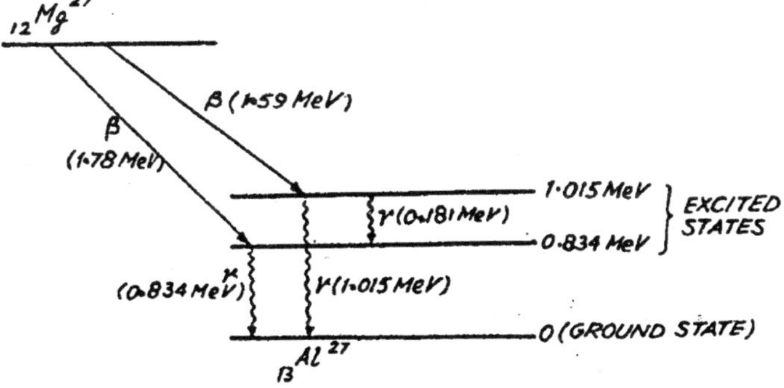

PROBLEMS

1. Find the kinetic energy of the α-particle emitted in the decay of $_{92}U^{232}$, assuming this atom to be at rest. The kinetic masses of U^{232}, Th^{228} and He^4 are 232.037168, 228.028750 and 4.002603 u respectively. Given: 1 u = 1.66 × 10^{-27} kg and 1 u × c^2 = 931.5 MeV.

Solution: The decay equation is

$$_{92}U^{232} \rightarrow {}_{90}Th^{228} + {}_2He^4$$

The Q-value (energy) for this decay is given by

$$\begin{aligned} Q &= [m\ (U^{232}) - m\ (Th^{228}) - m\ (He^4)]\ c^2 \\ &= (232.037168\ u - 228.028750\ u - 4.002603\ u)\ c^2 \\ &= 0.005815\ u \times c^2. \end{aligned}$$

Now,

$$1\ u \times c^2 = 931.5 \text{ MeV}.$$

$$\therefore \qquad Q = 0.005815 \times 931.5 = 5.42 \text{ MeV}.$$

Since Q-value is positive, the energy is *liberated* and so the decay occurs.

The liberated energy of 5.42 MeV is distributed between the α-particle and recoiling daughter nucleus of $_{90}Th^{228}$, that is,

$$\frac{1}{2} m\ (Th^{228})\ vR^2 + \frac{1}{2}\ m\ (He^4)\ v_a^{\ 2} = 5.42 \text{ MeV}, \qquad \dots (i)$$

where v_R and v_a are the velocities of the recoiling nucleus and the α-particle respectively.

The actual distribution can be calculated by linear momentum conservation, which gives

$$0 = m\ (Th^{228})\ v_R - m\ (He^4)\ v_a.$$

The nucleus of U^{232} is initially at rest, and the Th^{228} and He^4 nuclei must move in opposite directions. Assuming the ratio of Masses approximately equal to the ratio of mass numbers, we can write from the last equation.

$$v_R = \frac{4}{228} v_\alpha.$$

Substituting this value of v_R in eq. (*i*), and writing it in *SI* units, we have

$$\frac{1}{2} \times (228 \times 1.66 \times 10^{-27} \text{ kg}) \times \left(\frac{4}{228} v_\alpha\right)^2 + \frac{1}{2} \times (4 \times 1.66 \times 10^{-27} \text{ kg}) \times v_\alpha^2$$

$$= 5.42 \times 1.6 \times 10^{-13} \text{ joule.}$$

Solving, we get

$$v_\alpha = 1.60 \times 10^7 \text{ m/s.}$$

$$\therefore \quad K_\alpha = \frac{1}{2} m (\text{He}^4) v_\alpha^2$$

$$= \frac{1}{2} \times (4 \times 1.66 \times 10^{-27} \text{ kg}) \times (1.60 \times 10^7 \text{ m/s})^2$$

$$= 8.50 \times 10^{-13} \text{ joule.}$$

$$= \frac{8.50 \times 10^{-13}}{1.6 \times 10^{-13}} = 5.31 \text{ MeV.}$$

2. ***$_{10}Ne^{23}$ decays to $_{11}Na^{23}$ by beta emission. What is the maximum kinetic energy of the emitted electrons ? The atomic masses of Ne^{23} and Na^{23} are 22.994466 u and 22.989770 u respectively.***

Solution: The decay equation is

$$_{10}\text{Ne}^{23} \rightarrow {}_{11}\text{Na}^{23} + {}_{-1}\beta^{\circ} + \bar{v}.$$

The *Q*-value of this decay is given by (ignoring electron mass)

$$Q = [m(Ne^{23}) - m(Na^{23})]\,c^2$$
$$= (22.994466 \text{ u} - 22.989770 \text{ u})\, c^2$$
$$= 0.004696 \text{ u} \times c^2$$
$$= 0.004696 \times 931.5 = 4.37 \text{ MeV}.$$

Except for a small correction for the kinetic energy of the recoiling Na^{23} nucleus, the *maximum* kinetic energy of the electrons is just equal to this.

3. Compute the minimum energy of a γ-ray photon which may produce an electron-positron pair. The rest mass of electron is 91 × 10⁻³¹ kg and the speed of light is 30 ×10⁸ m/s.

Or

Why does a photon of wavelength 1 Å cannot produce an electron-positron pair ?

Solution: When an energetic γ-rays photon falls on a heavy substance, it is absorbed by some nucleus of the substance and an electron and a positron are produced. This phenomenon is called 'pair-production, and may be represented by the following equation:

$$\underset{(\gamma\text{-photon})}{h\nu} = \underset{(\text{positron})}{{}_{+1}\beta^0} + \underset{(\text{electron})}{{}_{-1}\beta^0}$$

According to Einstein's mass-energy relation, every body in the state of rest has some energy, called its rest-mass energy. If the rest-mass of the body be m_0, then its rest-mass energy is

$$E_0 = m_0c^2.$$

The rest-mass of each of the electron and the positron is 9.1×10^{-31} kg. So, rest-mass energy of each of them is

$$E_0 = m_0c^2$$
$$= (9.1 \times 10^{-31} \text{ kg}) \times (3.0 \times 10^8 \text{ m/s})^2$$
$$= 8.2 \times 10^{-14} \text{ joule}$$

$$= \frac{8.2 \times 10^{-14}}{1.6 \times 10^{-13}} = 0.51 \text{ MeV.}$$

[because 1 MeV = 1.6×10^{-13} joule]

Hence, for pair-production, it is essential that the energy of γ-photon must be at least $2 \times 0.51 = 1.02$ MeV.

The wavelength corresponding to this minimum photon-energy is

$$\lambda = \frac{hc}{E} = \frac{(6.62 \times 10^{-34} \text{ joule-sec}) \times (3.0 \times 10^{8} \text{ meter/sec})}{(1.02 \times 1.6 \times 10^{-13} \text{ joule})}$$

$$= 1.2 \times 10^{-12} \text{ meter} = 0.012 \text{ Å}$$

This is the *maximum* wavelength of the photon to produce electron-positron pair. Hence a 1.Å photon cannot produce the pair.

3

Radiation Absorption

The Laser

The word 'LASER' is an acronym for Light Amplification by Stimulated Emission of Radiation. It is a device to produce a strong, monochromatic, collimated and highly coherent beam of light; and depends on the phenomenon of "stimulated emission", first predicted by Einstein in 1916. Einstein considered the equilibrium between matter and electromagnetic radiation in a black-body chamber at a constant temperature in which exchange of energy takes place due to absorption and spontaneous emission of radiation by the atoms. He found that the usual absorption and emission processes alone are not sufficient to explain the equilibrium. He then predicted that there must be a third process also, now called "stimulated emission". This prediction was paid little attention until 1954, when Townes and Gordon developed a microwave amplifier (MASER) using ammonia, NH_3. In 1958, Schawlow and Townes showed that the maser principle could be extended into the visible region and in 1960, Maiman built the first laser using ruby as the active medium. Since then laser has opened up completely new fields of development in optics.

Radiative Absorption

An atom has a number of possible quantised energy states characterised by integral numbers. If it is initially in a lower state 1, it can rise to a higher state 2 by absorbing a quantum of radiation (photon) of frequency v, given by

$$v = \frac{E_2 - E_1}{h},$$

where E_1 and E_2 are the energies of the atom in the states 1 and 2 respectively (Fig.). This is absorption of radiation. This is a stimulated (or induced) process, the absorbed photon being the stimulating photon. (The absorption is necessarily stimulated).

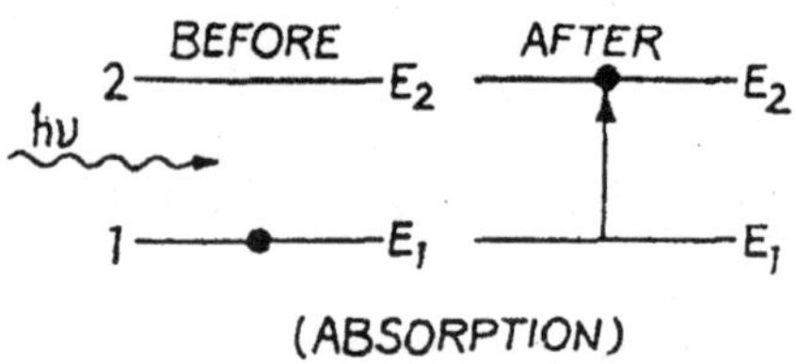

(ABSORPTION)

The probable rate of occurrence of this absorption transition 1 → 2 depends on the properties of states 1 and 2 and is proportional to the energy density $u(v)$ of the radiation of frequency v incident on the atom. Thus

$$P_{12} = B_{12}\, u(v). \qquad \text{... } (i)$$

The proportionality constant B_{12} is known as 'Einstein's coefficient of absorption of radiation'.

Natural Emission

Let us now consider an atom initially in the higher (excited) state 2 (Fig.). Observations show that its life-time in higher state is usually very small C $\approx 10^{-8}$ second) and it, of its own accord, jumps to the lower energy state 1, emitting a photon of frequency v. This is 'spontaneous' emission of radiation. If there is an assembly of atoms, the radiation emitted spontaneously by each atom has a random direction and a random phase, and is therefore incoherent from one atom to another.

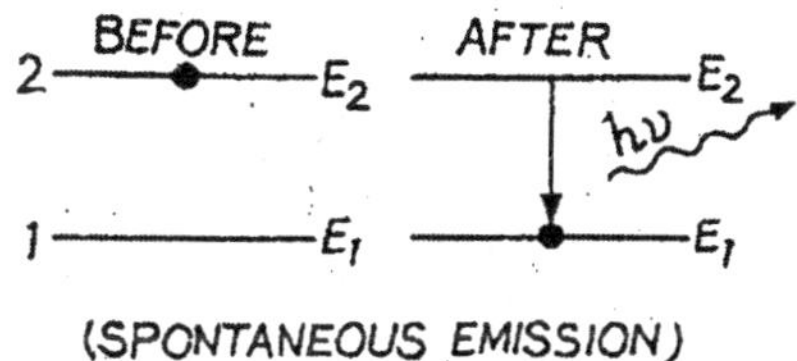

(SPONTANEOUS EMISSION)

The probability of spontaneous emission $2 \rightarrow 1$ is determined only by the properties of states 2 and 1. Einstein denoted this probability per unit time by

$$A_{21}$$

which is known as 'Einstein's coefficient of spontaneous emission of radiation'.

We note that the probability of absorption transitions depends upon the energy density ↔(v) of the incident radiation, whereas the spontaneous emissions are independent of it. Hence, for equilibrium, emission transitions depending upon $u(v)$ must also exist. These are 'stimulated' emission transitions.

Stimulated (or Induced) Emission: According to Einstein, an atom in an excited energy state may, under the influence of the electromagnetic field of a photon of frequency v incident upon it, jump to a lower energy state, emitting an additional photon of same frequency v (Fig.). Thus now two photons, one original and the other emitted, move on. This is 'stimulated' emission of radiation (or negative absorption of radiation). The direction of propagation, energy, phase and state of polarisation of the emitted photon is exactly the same as that of the incident stimulating photon. In other words, the stimulated radiation is completely coherent with the stimulating radiation. As a result of this process, radiation passing through an assembly of atoms is amplified.

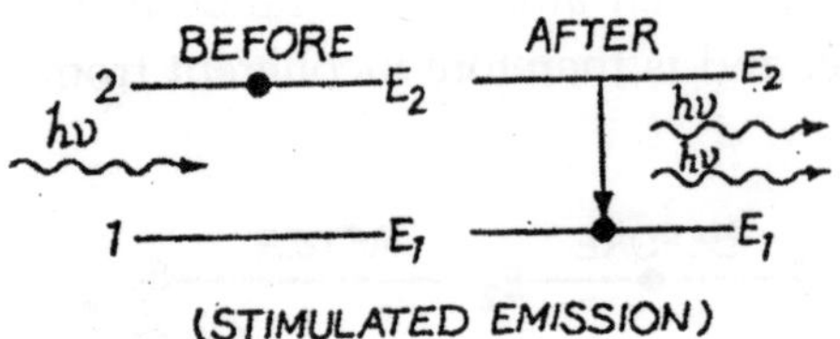

(STIMULATED EMISSION)

The probability of stimulated emission transition 2 → 1 is proportional to the energy density $u(v)$ of the stimulating radiation and is written as

$$B_{21}\ u(v),$$

where B_{21} is the 'Einstein's coefficient of stimulated emission of radiation'.

The total probability for an atom in state 2 to drop to the lower state 1 is therefore

$$P_{21} = A_{21} + B_{21}\ u(v). \qquad \text{... (ii)}$$

Relation between Spontaneous and Stimulated Emission Probabilities: Let us consider an assembly of atoms in thermal equilibrium at temperature T with radiation of frequency v and energy density $u(v)$. Let N_1 and N_2 be the number of atoms in states 1 and 2 respectively at any instant. The number of atoms in state 1 that absorb a photon and rise to state 2 per unit time is

$$N_1P_{12} = N_1B_{12}\ u(v). \qquad \text{[by eq. (i)]}$$

Conversely, the number of atoms in state 2 that drop to 1, either spontaneously or under stimulation, emitting a photon per unit time is

$$N_2P_{21} = N_2\ [A_{21} + B_{21}\ u(v)] \qquad \text{[by eq. (ii)]}$$

For equilibrium, the absorption and emission must occur equally.

Thus

$$N_1P_{12} = N_2P_{21}$$

or $$N_1B_{12}\, u(v) = N_2\, [A_{21} + B_{21}\, u(v)$$

or $$u(v) = \frac{N_2A_{21}}{N_1B_{12} - N_2B_{21}}$$

or $$u\,(v) = \frac{A_{21}}{B_{21}} \frac{1}{\frac{N^1}{N_2}\left(\frac{B_{12}}{B_{21}}\right) - 1}$$

Einstein proved thermodynamically that the probability of (stimulated) absorption is equal to the probability of stimulated emission, *i.e.*

$$B_{12} = B_{21}$$

Then, we have

$$u\,(v) = \frac{A_{21}}{B_{21}} \frac{1}{\frac{N_1}{N_2} - 1}$$

The equilibrium distribution of atoms among different energy states is given by Boltzmann's law according to which

$$\frac{N_2}{N_1} = \frac{e^{-E_2/kT}}{e^{-E_1/kT}}$$

or $$\frac{N_1}{N_2} = e^{-(E_2 - E_1)/kT} = e^{hv/kT}$$

Consequently, $$u(v) = \frac{A_{21}}{B_{21}} \frac{1}{e^{hv/kT}{}_{-1}}$$

This is a formula for the energy density of photon of frequency v in equilibrium with atoms in energy states 1 and 2, at temperature *T*. Comparing it with the Planck radiation formula

$$u\,(v) = \frac{8\pi hv^3}{c^3} \frac{1}{e^{hv.kT}{}_{-1}},$$

we get $$\boxed{\frac{A_{21}}{B_{21}} = \frac{8\pi hv^3}{c^3}}$$

This is the formula for the ratio between the spontaneous emission, and induced emission coefficients. This ratio is proportional to v^3. It means that the probability of spontaneous emission increases rapidly with the energy difference between two states.

Trigerred Radiation: Traits

Let us consider an ensemble of atoms irradiated with light of frequency v which coincides with one of the characteristic frequencies of the atoms. At room temperature, or higher, a certain number of atoms will be in an excited state. Now, two processes may occur: (*i*) an absorption transition of atoms from a lower energy level 1 to a higher energy level 2, such that $v = (E_2 - E_{1,})/h$; (*ii*) stimulated emission transition from higher energy level 2 to lower energy level 1.

In the first process, a photon from the incident beam is absorbed by an atom, thus leading the atom to an excited state. In the second process, an incident photon forces the excited atom to emit another photon of the same frequency in the same direction and in the same phase. The two photons go off together as coherent radiation.

Requirements for Laser Action: Under ordinary conditions of thermal equilibrium the number of atoms in higher energy state 2 is considerably smaller than the number in lower energy state 1 ($N_2 < N_1$), so that there is very little stimulated emission compared with absorption. An incident photon is more likely to be absorbed than to cause emission.

If, however, by some means a larger number of atoms are made available in the higher energy state, stimulated emission is promoted. The situation in which the number of atoms in the higher energy state exceeds that in the lower state ($N_2 > N_1$) is known as "population inversion". In this situation the system of atoms would lase.

Stimulated emissions are further encouraged by increasing

the radiation density $u(v)$ of the stimulating radiation. This is achieved by enclosing the emitted radiation in a "cavity" between two parallel reflectors. The radiation repeatedly travels back and forth, and the photons passing through the atoms go on multiplying by repeated stimulated emission (Fig.). Hence a strong coherent beam of light emerges from the system. (Fig.)

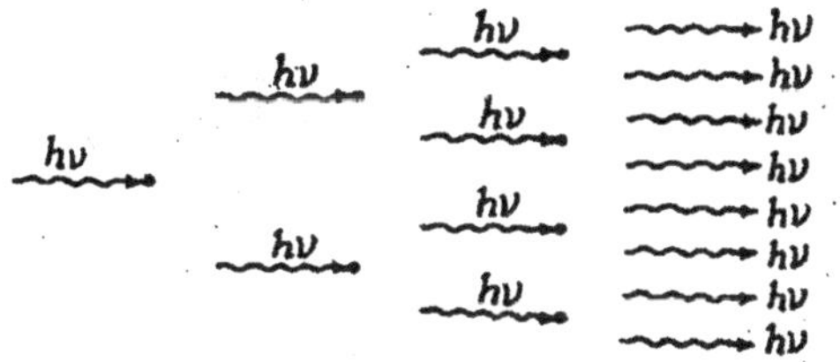

Pumping: The process of achieving population inversion is known as "pumping" of atoms. There are various types of pumping process, but the most natural is the 'optical pumping' which is utilised in Ruby laser.

The Ruby Laser: This is the first laser developed in 1960, and is a solid-state laser. It consists of a pink ruby cylindrical rod whose ends are optically flat and parallel (Fig.). One end is fully silvered and the other is only partially silvered. Upon the rod is wound a coiled flash lamp filled with xenon gas.

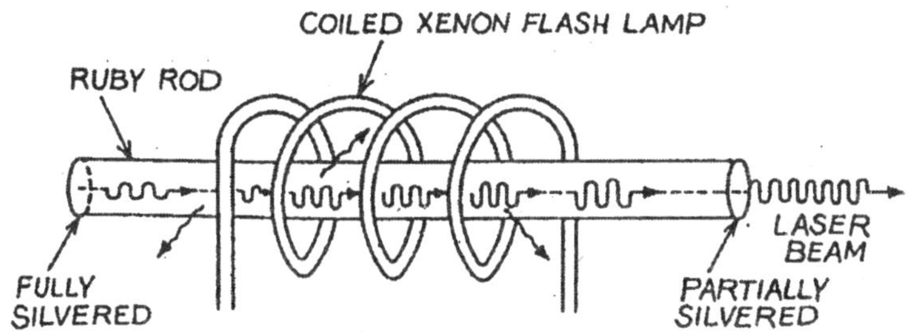

Working: The ruby rod is a crystal of aluminium oxide (Al_2O_3) doped with 0.05% cromium oxide (Cr_2O_3), so that some of the $A1^{+++}$ ions are replaced by Cr^{+++} ions. These "impurity" chromium ions give pink colour to the ruby and give rise to the laser action.

In Fig. is shown a simplified version of the energy-level diagram of chromium ion. It consists of an upper short-lived energy level (rather energy band) E_3 above its ground-state energy level E_1, the energy difference $E_3 - E_1$ corresponding to a wavelength of about 5500 A. There is an intermediate excited-state level E_2 which is metastable having a life-time of 3×10^{-3} sec (about 10^5 times greater than the life-time of E_3 which is $\approx 10^{-8}$ sec).

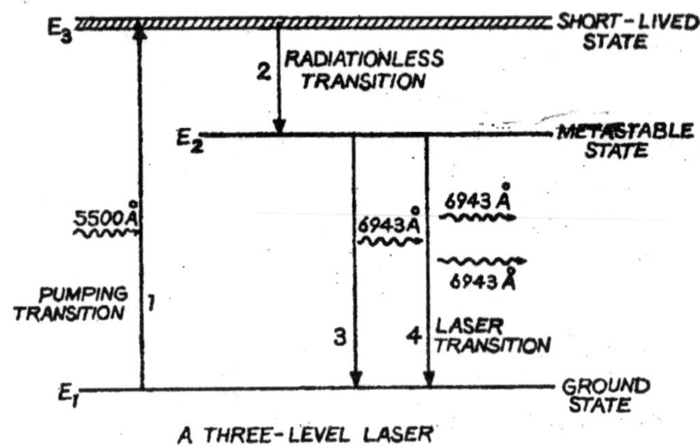

A THREE-LEVEL LASER

Normally, most of the chromium ions are in the ground state E_1. When a flash of light (which lasts only for about a millisecond) falls upon the ruby rod, the 5500-Å radiation photons are absorbed by the cromium ions which are "pumped" (raised) to the excited state E_3. The transition 1 is the (optical) pumping transition.

The excited ions give up, by collision, part of their energy to the crystal lattice and decay to the "metastable" state E_2. The corresponding transition 2 is thus a radiationless transition. Since the state E_2 has a much longer life-time, the number of ions in this state goes on increasing while, due to pumping, the number in the ground state E_1 goes on decreasing. Thus population inversion is established between the metastable (excited) state E_2 and the ground state E_1.

When an (excited) ion passes spontaneously from the metastable state to the ground state (transition 3), it emits a photon of wavelength 6943 Å. This photon travels through the ruby rod and, if it is moving parallel to the axis of the crystal, is reflected back and forth by the silvered ends until it stimulates an excited ion and causes it to emit a fresh photon in phase with the stimulating photon. This "stimulated" transition 4 is the laser transition. (The photons emitted spontaneously which do not move axially escape through the sides of the crystal).

The process is repeated again and again because the photons repeatedly move along the crystal being reflected from its ends. The photons thus multiply. When the photon-beam becomes sufficiently intense, part of it emerges through the partially-silvered end of the crystal.

There is a drawback in the three-level laser such as ruby. The laser requires high pumping power because the laser transition terminates at the ground state and more than one-half of the ground-state atoms must be pumped up to the higher state to achieve population inversion. Moreover, ions which happen to be in their ground state absorbs the 6943-Å photons from the beam as it builds up.

The ruby laser is a "pulsed" laser. The active medium (Cr^{+++} ions) is excited in pulses, and it emits laser light in pulses. While the Xenon pulse is of several millisecond duration; the laser pulse is much shorter, less than a millisecond duration. It means enhanced instantaneous power.

Helium-Neon Laser: It is a four-level laser in which the population inversion is achieved by electric discharge. A mixture of about 7 parts of helium and 1 part of neon is contained in a glass tube at a pressure of about 1 mm of mercury. (Fig.). At both ends of the tube are fitted optically plane and parallel mirrors, one of them being only partially silvered. The spacing of the mirrors is equal to an integral number of half-wavelengths of the laser light. An electric discharge is produced in the gas-mixture by electrodes connected to a high-frequency electric source.

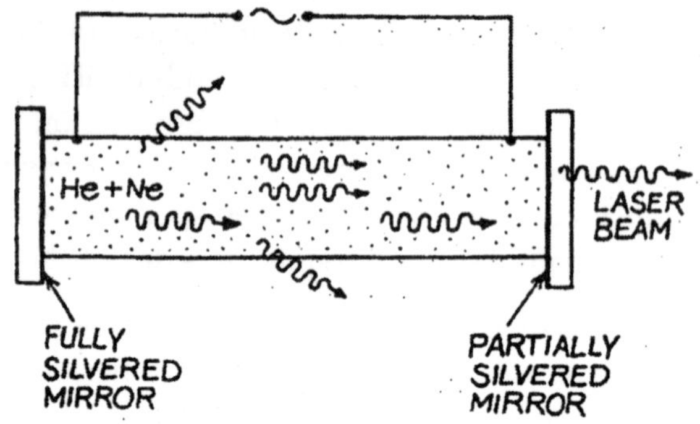

The electrons from the discharge collide with and "pump" (excite) the He and Ne atoms to metastable states 20.61 eV and 20.66 eV respectively above their ground states (Fig.). Some of the excited He atoms transfer their energy to ground-state Ne atoms by collisions, with the 0.05 eV of additional energy being provided by the kinetic energy of atoms. Thus He atoms help in achieving a population inversion in the Ne atoms.

When an excited Ne atom passes spontaneously from the metastable state at 20.66 eV to state at 18.70 eV, it emits a 6328-019Å photon. This photon travels through the gas-mixture, and if it is moving parallel to the axis of the tube, is reflected back and forth by the mirror-ends until it stimulates an excited Ne atom and causes it to emit a fresh 6328-Å photon in phase with the stimulating photon. This stimulated transition from 20.66-eV level to 18.70-eV level is the laser transition. This process is continued and a beam of coherent radiation builds up in the tube. When this beam becomes sufficiently intense, a portion of it escapes through the partially-silvered end.

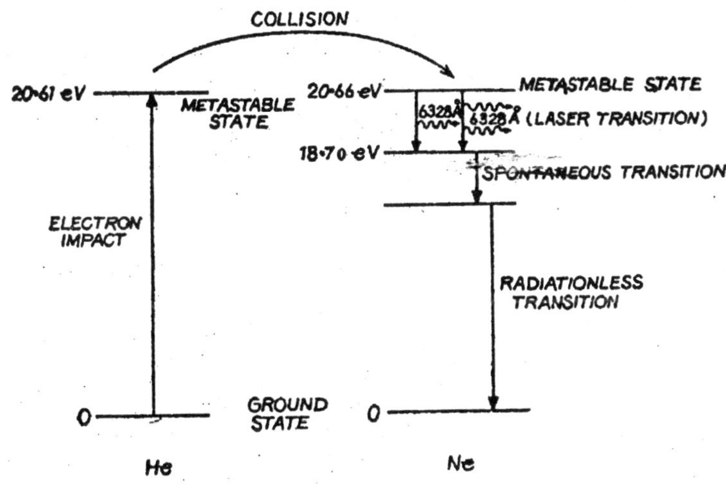

A FOUR-LEVEL LASER

From the 18.70-eV level the Ne atom passes down spontaneously to a lower metastable state emitting incoherent light, and finally to the ground state through collision with the tube walls. The final transition is thus radiationless.

Obviously, the Ne, atom in its ground state cannot absorb the 6328-A photons from the laser beam, as happens in the three-level ruby laser. Also, because the electron impacts that exite the He and Ne atoms occur all the time, unlike the pulsed excitation from the xenon flash lamp in the ruby laser, the He-Ne laser operates continuously.

Further, since the laser transition does *not* terminate at the ground state, the power needed for excitation is less than that in a three-level laser.

Properties of a Laser Beam: The laser beam has certain characteristic properties which are not present in beams derived from other light sources:

(i) The laser beam is completely spatially coherent, with the waves all exactly in phase with one another. An interference pattern can be obtained not merely by placing two slits in a laser beam but also by using beams from separate lasers.

(ii) The laser light is almost perfectly monochromatic, *i.e.* highly temporally coherent.

(iii) The laser rays are almost perfectly parallel. Hence a laser beam is very narrow and can travel to long distances without spreading. It can be brought to an extremely sharp focus.

(iv) The laser beam is extremely intense. It can vaporise even the hardest metal. Because of its high energy density and directional property, a laser beam can produce temperatures of the order of 10^4°C at a focussed point.

Applications of Lasers: The laser beam being narrow, intense, parallel, monochromatic and highly coherent is finding increasing applications in various fields:

(i) In the technical and industrial field, the laser beam is used for cutting fabric for clothing on one hand and steel sheets on the other. It can drill extremely fine holes in paper clips, single human hair and hard materials including teeth and diamond. Extremely thin wires used in cables are drawn through the diamond hole. Metallic rods can be melted and joined by means of a laser beam (laser welding). The surfaces of engine crankshafts and the cylinder walls are hardened through heat-treatment by laser. The laser beam is used to vaporise unwanted material during the manufacture of electronic circuits on semiconductor chips.

(ii) In the medical field, the laser beam is used in delicate surgery like cornea grafting. Using laser beam, the surgical operation is completed in a much shorter time. It is also used in the treatment of kidney stone, cancer, tumour and in cutting and sealing the small blood vessels in brain operation.

(iii) During war-time, lasers are used to detect and destroy enemy missiles. Now, laser-rifles, laser-pistols and laser-bombs are also being made which can be aimed at the enemy in the night. In space, laser has been used to control rockets and satellites and in directional radio-communication like fiberoptic telephony.

(iv) Laser is very useful in science and research. It has been used to perform Michelson-Morley experiment which is the building stone of the Einstein's theory of relativity. It can be used to determine the temperature of plasma and the density of electron. Laser-torch is used to see objects at long distances.

(v) Laser is used in holography and nonlinear optics.

(vi) Since laser rays are very much parallel, so they are used for communications and measuring long distances. The distance between earth and moon has been measured by laser rays to an accuracy of 15 cm.

(vii) Laser rays have proved to be useful in detecting nuclear explosions and earthquakes, in vaporising solid fuel of rockets, in the study of the surface of distant planets and satellites.

(viii) Laser beams have also been used in the "inertial confinement" of plasma.

4

Nuclear Disintegration

Artificial Radioactivity

Nuclear Fission

Rutherford's Discovery of Artificial—Nuclear Disintegration—Discovery of Proton: The radioactive elements, in which natural disintegration of nucleus occurs, are the heavier elements. Rutherford was working for artificially disintegrating the nuclei of *lighter* elements by bombarding them with high-speed particles. He, in 1919, succeeded in disintegrating nitrogen nuclei by bombarding ordinary nitrogen gas with α-particles emitted from RaC′.

Rutherford's apparatus consisted of a long chamber with a side opening covered by a silver foil F (Fig.). A zinc sulphide (ZnS) screen was placed just outside the opening and a microscope M was placed for observing any scintillations occurring on the screen. The source of α-particles, S, was deposited on a metal plate placed inside the chamber. The distance of S from the ZnS screen could be varied. The chamber could be filled with different gases through the side-tubes T_1 and T_2.

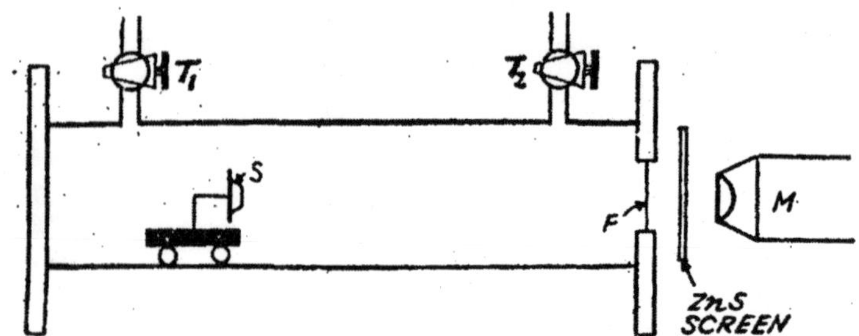

When the chamber was filled with oxygen or carbon-dioxide gas, no scintillations were seen on the screen. *But when it was, felled with nitrogen, scintillations were observed on the screen,* even though this screen was shielded from the α-particle source S by the silver foil *F* which was thick enough to absorb all the α-particles. Hence Rutherford concluded that the scintillations were produced by some new particles which were more penetrating that the α-particles and were ejected from the nitrogen nuclei by collision with α-particles.

Since the scintillations were observable even when the distance of *S* from the screen was as long as 40 cm, the ejected particles had a range upto 40 cm. Magnetic deflection experiments indicated that these particles had a positive charge equal to the charge of an electron and a mass equal to that of a hydrogen nucleus (1.672×10^{27}kg). They were called as 'protons'.

Blackett, in 1925, took cloud chamber photographs of the above process. He obtained a straight thicker track of the incident α-particle, and a long thinner track of the ejected proton together with a short, thick track of the recoiling nucleus. *There was no α-particle track after the collision, showing* that the α-particle had disappeared completely. From this it was concluded that the α-particle did not rebound after collision but was absorbed by the nitrogen nucleus, resulting into a new nucleus and the emission of a proton. The nuclear reaction for this process is

$$\underset{\text{nitrogen}}{_7N^{14}} + \underset{\alpha\text{-particle}}{_2He^4} \rightarrow \underset{\text{oxygen isotope}}{_8O^{17}} + \underset{\text{proton}}{_1H^1}$$

This reaction states that when a nitrogen nucleus ($_7N^{14}$) is hit by an α-particle ($_2He^4$), a proton ($_1H^1$) is ejected leaving a recoiling oxygen nucleus ($_8O^{17}$). Thus Rutherford transformed ordinary nitrogen into a rare isotope of oxygen. This was the first *artificial* nuclear transformation.

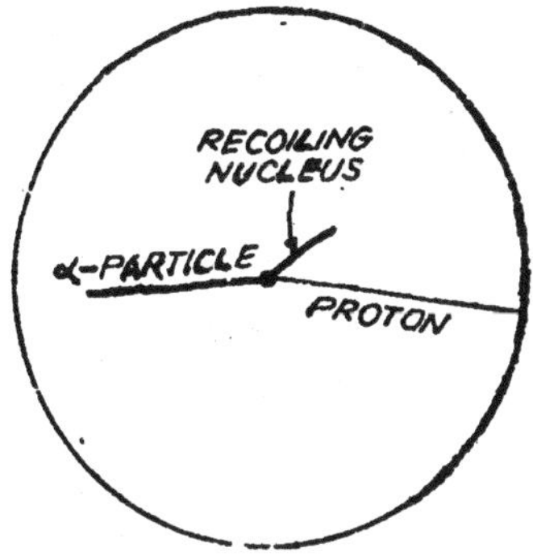

Following Rutherford historic experiment on artificial nuclear disintegration, Rutherford and Chadwick disintegrated other light elements by bombarding them with α-particles. They found that protons could be ejected out of the nuclei of all the light elements from boron to potassium (with the exception of carbon and oxygen).

In some cases the energy of the ejected protons was even greater than that of the bombarding α-particles. This result further verified that the protons were emitted due to nuclear disintegration, the extra energy being acquired due to the nuclear re-arrangement.

Later on, particles other than those emitted from a radio-active substance, such as neutrons, and artificially-accelerated charged particles (protons, deuterons) were used for producing nuclear disintegration.

Neutron Found Out

In 1920, there were only three known material particles, namely electron, proton and α-particle, all charged. Rutherford had suggested the existence of a new particle having a mass roughly equal to the proton mass but carrying no charge. The particle was subsequently discovered and is now known as the neutron.

In 1930, Bothe and Becker found that when beryllium (or boron) was bombarded with α-particles, a highly penetrating but very poorly ionizing radiation was emitted. It was supposed that this radiation was high-energy γ–radiation produced by the reaction

$$_4Be^9 + {}_2He^4 \rightarrow {}_6C^{13} + \gamma$$

This supposition led, however, to difficulties. The measurement of absorption of the radiation in lead showed that if it was a γ–radiation then its energy should be about 7 MeV. This value was greater than the energy of any γ–radiation known at that time.

In 1932, Curie and Joliot investigated the radiation further by examining its effect on a paraffin block (a substance rich in hydrogen) placed between the beryllium and an ionisation chamber (Fig.). They found that the ionisation increased markedly by the presence of the block. They correctly concluded that protons (hydrogen nuclei) were being ejected from the paraffin by the radiation, and producing ionisation in the chamber. They assumed that the supposed γ–radiation on falling upon the paraffin, underwent Compton collisions with the hydrogen nuclei, which therefore recoiled and appeared as protons.

These protons were found to have energies of about 4.5 MeV. Calculations on this basis showed that each incident γ–ray photon must have had an energy of 55 MeV, a value about eight times higher than that deduced from the absorption measurements. Thus there was a serious difference between the values of the energy of the supposed γ–radiation given by the two methods.

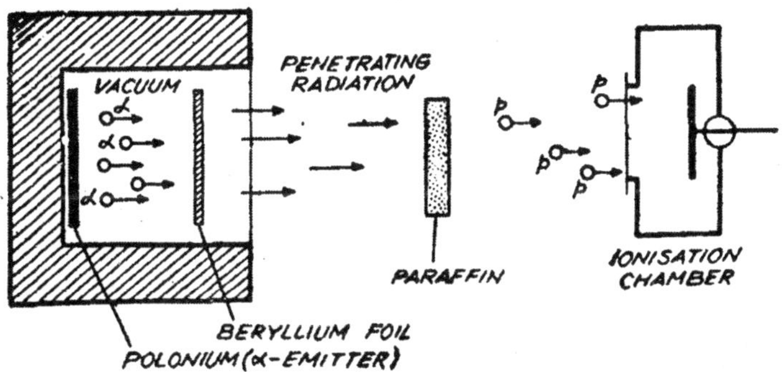

In 1932, Chadwick performed a series of experiments on the recoil of many other nuclei (for example nitrogen) when struck by this penetrating radiation. He found that if this radiation consisted of γ–ray photons, then the energy of the photons as obtained from experimental results varied with the nature of the recoiled nucleus. For example, protons ejected from paraffin had energies which required a γ–ray photon to have an energy of 55 MeV, while recoiling nitrogen nuclei had energies which required a γ–ray photon to have an energy of 90 MeV. This meant that the energy of the supposed γ–ray photon increased with the mass of the recoiling atom, which was contrary to the conservation of energy and momentum in Compton collisions.

Chadwick showed that these difficulties disappeared if the radiation coming from beryllium bombarded with α-particles is supposed to consist of 'particles' (instead of massless γ–ray photons) of mass nearly equal to that of a proton, but having no charge. He called these particles 'neutrons'. The nuclear reaction which produces these neutrons is

$${}_4Be^9 + {}_2He^4 \rightarrow {}_6C^{12} + {}_0n^1$$

where ${}_0n^1$ is the symbol for the neutron. The penetrating nature of the neutrons follows from the absence of a charge, and the energies of the recoiling nuclei in Chadwick's experiments can be accounted for on the basis of collisions with an energetic 'particle.'

Production of Neutrons: As we have seen above, neutrons are produced when a light element like beryllium (or boron) is bombarded with α-particles. Therefore an α-emitter (like radium, polonium or americium) mixed with beryllium powder constitutes a neutron source. Such a mixture placed in a capsule emits neutrons with energies ranging upto 10 MeV or more.

Neutrons can also be produced by bombarding deuterium (heavy water) or beryllium with γ–rays obtained from artificial radioactive atoms like ${}_{11}Na^{24}$ or ${}_{51}SB^{124}$. Such sources are called 'photoneutron' sources and give practically monoenergetic neutrons.

Now a days the most powerful source of neutrons is the nuclear reactor in which the fission of heavy nuclei takes place.

Determination of Mass of Neutron: Since neutron is not a charged particle, its mass cannot be determined directly by deflecting it in electric or magnetic field. Chadwick determined the mass of neutron by measuring the maximum velocities of recoiling nuclei of hydrogen and nitrogen struck by neutrons. Suppose a neutron of mass m and velocity v suffers head-on collision with a stationary hydrogen nucleus of mass m_H. Let the velocity of the neutron after the collision be v', and the (maximum) velocity of the recoiling hydrogen nucleus be v_H. The equation of conservation of kinetic energy is

$$\frac{1}{2}mv^2 = \frac{1}{2}mv'^2 + \frac{1}{2}m_{H^vH^2},$$

and the equation of conservation of momentum is

$$mv = mv' + m_H v_H.$$

Eliminating v' from these two equations, we get

$$v_H = \frac{2m}{m + m_H} v. \qquad \text{... } (i)$$

Again, if a neutron with the same velocity v collides with a stationary nitrogen nucleus of mass m_N, the (maximum) velocity imparted to the recoiling nitrogen nucleus is given by

$$v_N = \frac{2m}{m + m_N} v. \qquad \text{... } (ii)$$

Dividing eq. (*i*) by (*ii*), we get

$$\frac{v_H}{v_N} = \frac{m + m_N}{m + m_H}$$

m_N and m_H are known. If, therefore, v_H and v_N are found by measuring the maximum length of the cloud chamber tracks of the recoiling nuclei, the mass m of the neutron can be determined. Chadwick's result, although approximate because of errors in the determination of v_H and v_N, showed that the mass of the neutron is slightly larger than that of the proton.

Properties:

(i) The neutron is a fundamental constituent of the nuclei of all atoms (except hydrogen atom). It has a mass of 1.00898 amu or 1.675×10^{-27} kg which is slightly greater than that of a proton.

(ii) It is an uncharged particle. Therefore it cannot be accelerated to high velocities by means of electric fields as can be charged particles such as protons and electrons. For the same reason, the neutrons cannot be focussed by means of magnetic fields.

(iii) It is a highly penetrating particle and can pass through thick sheets of lead.

(iv) Being chargeless, the neutron produces practically no ionisation in a gas, and hence no track in a cloud chamber.

(v) Being uncharged, a neutron can easily enter the nucleus of an atom. It can therefore produce a nuclear excitation or nuclear disintegration far more readily than almost any other particle. (Other particles carrying a charge have to overcome the strong electrostatic repulsion offered by the nucleus.)

(vi) If the probability of nuclear excitation or nuclear disintegration is small, the neutrons, on striking matter, are simply scattered by the atomic nuclei. On colliding with heavy nuclei, the neutrons are scattered with very little loss of energy. On colliding with light nuclei, however, the neutrons are slowed down in a few collisions. Light water (H_2O), heavy water (D_2O), paraffin wax, and carbon are very effective in slowing down neutrons. These substances are called 'moderators'. A slow neutron is more efficient in producing nuclear disintegration because it spends more time near a nucleus than a fast neutron and thus stands a greater chance of being captured.

(vii) Neutrons possessing energies of 1 MeV or more are known as *fast* neutrons. Those with energies below 1 eV are described as *slow* neutrons. The neutrons which have come into thermal equilibrium with a moderator at normal temperature and pressure are called *thermal* neutrons. Such neutrons have energies of approximately 0.03 eV.

Detection of Neutrons: The neutron, being a non-ionising particle, does not produce a track in a cloud chamber. Hence it cannot be detected by a G-M tube or by a cloud chamber. Indirect methods are, however, available for its detection.

Slow neutrons can be detected by means of a G-M tube or ionisation chamber filled with boron trifluoride gas (containing ${}_5B^{10}$). The neutrons passing through the gas disintegrate boron nuclei which thus emit α-particles:

$$ {}_5B^{10} + {}_0n^1 \rightarrow {}_8Li^7 + {}_2He^4 \text{ (}\alpha\text{-particle)} $$

The α-particles so produced cause ionisation of the gas and are thus detected.

Similarly, a G-M tube or ionisation chamber containing hydrogen will detect fast neutrons. When a fast neutron collides with a hydrogen nucleus, it imparts energy to the hydrogen nucleus which in turn produces ionisation in the surrounding gas and is detected.

Radioactive Decay of Neutron: A free neutron outside an atomic nucleus is *unstable* and decays into a proton, emitting an electron and an antineutrino. The reaction is

$$ {}_0n^1 \rightarrow {}_1H^1 + {}_{-1}e^0 + \bar{v} $$

(neutron) (proton) (electron)
(antineutrino)

The half-life of the neutron has been estimated to be 12.8 min.

Uses of Neutrons:

(i) Neutrons are used in medicine, specially in the treatment of cancer.

(ii) Fast and slow neutrons are used for artificial disintegration of nuclei and producing radio-isotopes.

(iii) Slow neutrons are used in nuclear fission.

Positron: It is a positively-charged particle having the same mass and charge as an (negative) electron. Thus it is the *antiparticle* of the electron and is also called as 'positive electron'.

Positron was discovered by Anderson in 1932. Anderson was photographing the tracks of comic-ray electrons in a Wilson cloud chamber placed in a strong, magnetic field. He obtained a number of curved tracks showing that they were formed by charged particles of electronic mass and electronic charge. The direction of curvature of most of the tracks indicated a negative charge on the particles. Occasionally, however, a track was obtained whose direction of curvature was just opposite, thus indicating a positive charge on the particle producing it. The particle was named positron. It was the first antiparticle to be discovered.

The existence of positron was predicted by Dirac before its discovery by Anderson. It was later found that when hard γ–rays are absorbed by nuclei of atoms, the electrons and positrons are simultaneously produced (pair production). Positrons are also emitted by some artificially produced radioactive substances.

A positron has only an extremely short life, of the order of a micro-second. It is readily annihilated by combining with an electron from the surroundings.

Artificial Transmutation of Elements (or Artificial Disintegration of Nuclei): An artificial disintegration of nuclei is a process in which a nucleus is transformed into a different species by its reaction with an energetic particle or photon. The formation of the product nucleus is accompanied by the emission of one or more light particles, or photon. The whole process occurs very rapidly in a time of about 10^{-13} second or less. It is also known as artificial nuclear disintegration' or 'nuclear reaction'. The nuclear reaction which was the first to be discovered by Rutherford is

$$\underset{}{{}_7N^{14}} + \underset{(\alpha\text{-particle})}{{}_2He^4} \rightarrow {}_8O^{17} + \underset{(\text{proton})}{{}_1H^1}$$

In this reaction a nitrogen nucleus hit by an α-particle is converted into an oxygen nucleus and emits a proton. For every nuclear reaction the total number of protons and neutrons is conserved as is the total mass *plus* energy.

Radioactivity not Real

The experiments on the artificial disintegration of nuclei led to the discovery of artificial (or induced) radioactivity.

In 1934, Curie and Jollot observed that when certain light elements, like boron and aluminum, were bombarded by α-particles, the resulting products of disintegration emitted positrons, and the emission persisted *even after bombardment by α-particles was stopped.*

The positron activity decayed exponentially with time just in the same way as natural radioactivity. This showed that as a result of α-particle bombardment, the stable elements were converted into unstable radioactive isotopes. *The phenomenon in which a stable element is converted into a radioactive isotope by an artificial disintegration is called 'artificial radioactivity'.* These artificially-produced radioactive isotopes have comparatively much shorter half-lives than the natural radioactive elements.

Thus when boron is bombarded by α-particles, it disintegrates emitting a neutron and forming an unstable radioactive isotope of nitrogen called radio-nitrogen:

$$_5B^{10} + {}_2He^4 \rightarrow {}_7N^{13} + {}_0n^1.$$

The radio-nitrogen $_7N^{13}$ decays into $_6C^{13}$ by emitting a positron:

$$_7N^{13} \rightarrow {}_6C^{13} + {}_{+1}e^0 \quad \text{(positron)}$$

The half-life of $_7N^{13}$ is found to be about 14 min, and $_6C^{13}$ produced is a stable isotope of carbon.

In a similar manner, the bombardment of aluminium results in the production of unstable radio-phosphorus, which decays by positron-emission into a stable isotope of silicon:

$$_{13}Al^{27} + {}_2He^4 \rightarrow {}_{15}P^{30} + {}_0n^1$$

$$_{15}P^{30} \rightarrow {}_{14}Si^{30} + {}_{+1}e^0$$

In general, artificial radioactive isotopes are produced by the bombardment of stable elements with accelerated charged particles (protons, deuterons, α-particles). By far the most useful particles for producing artificial radioactivity are the neutrons which are available in large numbers in a nuclear reactor. Hence artificially radioactive isotopes are now-a-days produced by placing elements inside a nuclear reactor. Practically all the elements can now be made artificially radioactive.

Some artificially radioactive isotopes emit positrons, while some others emit electrons. In some cases γ–ray emission also takes place together with positrons or electrons. Artificially radioactive isotopes which emit α-particles are much less common and occur mostly in transuranic elements (Z>92).

There is some correlation between the type of radioactivity of an artificially radioactive isotope and the means of its formation. The radioactive isotopes produced by (n, γ), (n, p), (n, α), and (d, p) reactions in which the neutron to proton ratio increases, emit electrons. As an example, we consider (n, α) reaction:

$$_{13}Al^{27} + {_0}n^1 \rightarrow {_{11}}Na^{24} + {_2}He^4$$

$$_{11}Na^{24} \rightarrow {_{12}}Mg^{24} + {_{-1}}e^{\circ} \text{ (electron)}$$

Radioactive isotopes produced by (p, γ), (p, n), (α, n), (d, n) and (γ, *n*) reactions in which the neutron to proton ratio decreases, emit positrons. As an example, we consider (α, n) reaction:

$$_7N^{14} + {_2}He^4 \rightarrow {_9}F^{17} + {_0}n^1$$

$$_9F^{17} \rightarrow {_8}O^{17} + {_{+1}}e^0 \text{ (positron)}$$

Radio-isotope and their Applications: In addition to the naturally occurring radio-isotopes such as radium, hundreds of others have been made artificially. These isotopes have numerous applications in medicine, agriculture, industry and pure research. Many applications employ a special technique known as 'tracer technique'.

Tracer Technique: A small quantity of a radio-isotope is introduced into the substance to be studied and its path is traced by means of a G. M. counter. As an example, a leakage in an underground water-pipe can be detected by this method. A small quantity of radio-sodium Na^{24} (γ–ray emitter) is introduced into the pipe at its inlet.

After the liquid has gone through the pipe, the ground around the leak will have larger γ–ray activity which can be detected by moving a G. M. counter on the ground. Similarly, in order to locate a blockage in an underground sewage pipe, a rubber ball having Na^{24} is introduced into the pipe. A G. M. counter above ground will give the position of the ball when it has come to rest.

Radio-isotopes can also be used in transporting *different* oils through underground pipe to distant places. When the type of the oil flowing through the pipe is changed, a small quantity of radio-isotope is mixed exactly at the position where the change takes place. Near the other end of the pipe line, a Geiger counter is placed which gives a signal when the radio-isotope passes.

In the field of medicine the tracer technique is employed in a number of ways. For example, the doctor can find out any obstruction in the circulation of the blood in the human body. He injects radio-phosphorous (P^{32}) into the blood of the patient and examines the movement of the blood by detecting radiations emitted by P^{32} by means of G. M. counter. He can thus locate clots of blood present in the body. In a similar way, the passage of a particular element in the body and the rate at which it accumulates in different organs can be studied. For example, phosphorus accumulates in bones, and iodine in thyroid gland. When the thyroid gland suffers with some disease, its rate of accumulation of iodine changes. To investigate it, radio-iodine (I^{131}) is given orally to the patient and the radiation emitted by his thyroid gland is measured externally by a G. M. counter at suitable intervals over the following 48 hours or so. An over-active or under-active (ailing) thyroid gland can thus be diagnosed.

In agriculture, the tracer technique is used to study the rate and direction of movement of an element in a plant. For this a radio-isotope of that element is injected in the ground near the plant. After a few days the plant is laid on a photographic paper to produce an auto-radiograph. The dark areas in the radiograph show the positions reached by the element. This technique gives valuable information regarding the optimum season for fertilising crops and for poisoning weeds.

In industry, the tracer technique is used for testing the uniformity of mixtures. For testing a chocolate mixture, a small quantity of a short-lived radio-isotope such as Na^{24} or Ma^{56} is added to the primary ingredients. Several different samples of the final product are then tested for radioactivity by means of a G. M. counter. If each sample gives the same counting rate, then the mixing has been uniform. This method can be used in mixing processes occurring in the manufacture of chocolate, soap, cement paints, fertilizers, cattle food and medical tablets.

The tracer technique is extremely sensitive in testing the sealing process in making envelopes for radio valves. A sample valve is filled with radio-krypton (Kr^{85}) and a G. M. counter is held outside the valve. The counter detects even an extremely poor leakage.

The tracer technique is also used in research to study the exchange of atoms between various molecules, and to investigate the solubility and vaporisation of materials.

Besides the uses employing the tracer technique there are hundreds of other uses of radio-isotopes in various fields.

Medical Uses: The radiations given out by some radio-isotopes are very effective in curing certain diseases. For example, radio-cobalt (Co^{80}) is used in the treatment of brain tumour, radio-phosphorus (P^{32}) in bone-diseases and radio-iodine (I^{131}) in thyroid cancer. The radiations, besides destroying the ailing tissue, also damage the healthy tissue and hence a careful control over the quantity administered is necessary.

Bacteria and other disease-carrying organisms can be destroyed by irradiating them with γ–rays. The process is used to sterilise medical instruments, plastic hypodermic needles, packets of antibiotics, and hospital blankets; whereas heat sterilisation would damage them. A portable source of γ–rays for sterilisation is radio-cobalt (Co^{60}).

X-ray photography in medical diagnosis can be replaced by γ–ray photography with advantage. The γ–ray source (radio-isotope) is compact and needs no power supply.

Agricultural Uses: Radiations from certain radio-isotopes are used for killing insects which damage the food grains. Certain seeds and canned food can be stored for longer periods by gently exposing them to radiations.

Better yields of milk from cows, and more eggs from hens have been obtained on the basis of information gained by mixing radio-isotopes with their diet.

Radio-isotopes are also employed for determining the function of fertilizer in different plants. Thus the agricultural yield is increased.

Certain seeds, when exposed to feeble radiation, develop into different varieties of plants. For example, new and exciting colours have been given to some of the flowering plants.

Industrial Uses: There are many different uses to which radio-isotopes are put in industry. By γ–ray photography we can find out wearing of cutting tools and lathes, and can locate internal cracks in stones. We can check any non-uniformity in the thickness of a sheet by β– or γ–absorption measurements.

The sheet is made to run continuously between a radio-isotope (emitting β– or γ–rays) and a counter. A change in the counting rate indicates a variation in the thickness of the sheet. The output from the counter may be used to correct the machinery which is rolling the sheet as soon as a variation is detected, and thus the thickness is automatically kept constant. This method is used as a thickness control in the manufacture of paper, plastic, metal sheet, etc.

The same method can be used to check sealed cigarette packets whether they are full or if one or more cigarette is missing. The packets are placed on a conveyer belt running between a radio-isotope and a counters. An empty or partially-filled packet gives a higher counting rate due to less absorption of radiation than with a completely filled packet. The increase in counting rate can be converted into an electronic signal which knocks the incomplete packet off the belt.

A radioactive isotope together with a fluorescent material (such as ZnS) becomes a weak source of light. Such sources are used for providing light in coal mines, and for painting watches.

Carbon Dating: The radio-carbon ($_6C^{14}$) is continuously being produced in the atmosphere by neutron bombardment of nitrogen:

$$_7N^{14} + {_0n^1} \rightarrow {_6C^{14}} + {_1H^1}$$

The radio-isotope $_6C^{14}$ has a half-life of 5600 years and decays back to nitrogen by electron emission:

$$_6C^{14} \rightarrow {_7N^{14}} + {_{-1}e^0}.$$

In the course of time an equilibrium has been reached in which the rates of formation and decay of $_6C^{14}$ are equal. Hence the ratio of $_6C^{14}$ to stable carbon $_6C^{12}$ in all atmospheric carbon dioxide is constant.

All *living* matter (plant or animal) by one process or another exchanges its carbon with atmospheric carbon (in the form of carbon dioxide). As long as the exchange is taking place, the ratio of ${}_6C^{14}$ to ${}_6C^{12}$ in the plant or animal is the same as that in the atmosphere. After the death of a plant or animal the process of exchange stops, so that the concentration of ${}_6C^{14}$ in it decreases due to radioactive decay. The ratio of ${}_6C^{14}$ to ${}_6C^{12}$ remaining at a certain time after death, can be found by measuring the activity of a carbon sample taken from the dead body. From this ratio the time since death can be calculated. By this method, ages ranging from about 600 years to 20,000 years can be estimated. The method is known as 'carbon-14 dating'.

Carbon-14 dating is applicable for dating dead vegetation (such as wood, paper, papyrus, etc.) and animal matter (such as hair, wool, etc.). For instance, it has been used for wooden coffins containing Egyptian mummies, which were found to be 3500 years old.

Transuranic Elements: The heaviest naturally occurring element is uranium (*U*) which has an atomic number Z = 92. Since 1940, elements having an atomic number greater than 92 have been *artificially* produced in the laboratory by the bombardment of certain heavy nuclei with appropriate particles. These are called 'transuranic elements' (Z>92). The transuranic elements ranging from Z=93 to Z=103 are the following:

(1) *Neptunium, Np (Z = 93):* When uranium ${}_{92}U^{238}$ is bombarded with *slow* neutrons, a new isotope ${}_{92}U^{239}$ is produced with the emission of γ–rays (radioactive capture):

$${}_{92}U^{238} + {}_0n^1 \rightarrow {}_{92}U^{239} + \gamma$$

The product nucleus ${}_{92}U^{239}$ is radioactive and decays by electron emission into ${}_{93}Np^{239}$, an isotope of the first transuranic element neptunium:

$${}_{92}U^{239} \rightarrow {}_{93}Np^{239} + {}_{-1}e^0 + \overline{v}$$

(2) *Plutonium, Pu (Z = 94):* The Neptunium isotope $_{93}Np^{239}$ is itself radioactive and has a half-life of 23 days. It decays by electron emission into $_{94}Pu^{239}$, an isotope of the second transuranic element plutonium:

$$_{98}Np^{239} \rightarrow {}_{94}Pu^{239} + {}_{-1}e^{0} + \overline{\upsilon}$$

$_{91}Pu^{239}$ is very stable. It decays by α-emission into $_{92}U^{235}$ with a half-life of 24,400 years:

$$_{94}Pu^{239} \rightarrow {}_{92}U^{235} + {}_{2}H_{e}^{4}$$

The production of $_{94}Pu^{239}$ is carried on a large scale for generating nuclear power.

Other transuranium elements can be made by bombarding the appropriate heavy nucleus with accelerated α-particles, deuterons, or with bare nuclei of lighter atoms. They are named as below:

(3) Americium, Am (Z = 95)

(4) Curium, Cm (Z = 96)

(5) Berkelium, Bk (Z = 97)

(6) Californium, Cf (Z=98)

(7) Einsteinium, E (Z = 99)

(8) Ferminm, Fm (Z=100)

(9) Mendelevich, Mv (Z = 101)

(10) Nobelium, No (Z=102)

(11) Lawrencium, Lw (Z = 103)

Each one of the transuranium elements is found to have several isotopes. Elements beyond *E* (Z=99) have so short half-lives that they cannot be isolated in weighable quantities.

In general, the half-lives of transuranium elements are very short compared with the age of the earth ($\approx 4 \times 10^9$ years). Even if the elements existed among the natural ores when the earth was formed, they would have disappeared long ago.

This is why they do not occur in nature. On the other hand, the half-life of uranium is about the same as the age of the earth so that a considerable quantity still remains.

PROBLEM

1. ***10 mg of carbon from living material produces 200 counts per minute due to a small fraction of the radio-carbon $_6C^{14}$. A piece of ancient wood of mass 10 mg is found to give 50 counts per minute. Find the age of the wood assuming that the $_6C^{14}$ content of the atmosphere has remained unchanged. Half-life of $_6C^{14}$ is 5700 years.***

[**Ans.** 11400 years]

5

Nuclear Fission

Radioactive Split

Nuclear Fission: In 1939, two German scientists Hahn and Strassman observed that when uranium nucleus (Z=92) was bombarded with neutrons, it splitted up into two radioactive nuclei which were identified as isotopes of barium (Z=56) and krypton (Z=36). Frisch and Meitner called this phenomenon as 'nuclear fission'.

Thus, *nuclear fission is a process in which a heavy nucleus, after capturing a neutron, splits up into two lighter nuclei of comparable masses.* The product nuclei are called 'fission fragments'. The process is accompanied by the release of a few fast neutrons and a huge amount of energy in the form of the kinetic energy of the fission fragments, and also as γ-rays (Fig.).

The natural uranium is a mixture of two isotopes, $_{92}U^{238}$ and $_{92}U^{235}$ in the ratio 145:1. The isotope $_{92}U^{238}$ can be fissioned only by the 'fast' neutrons having energy above 1 MeV, while $_{92}U^{235}$ can be fissioned by quite slow neutrons also, say by thermal (0.03 eV-energy) neutrons.

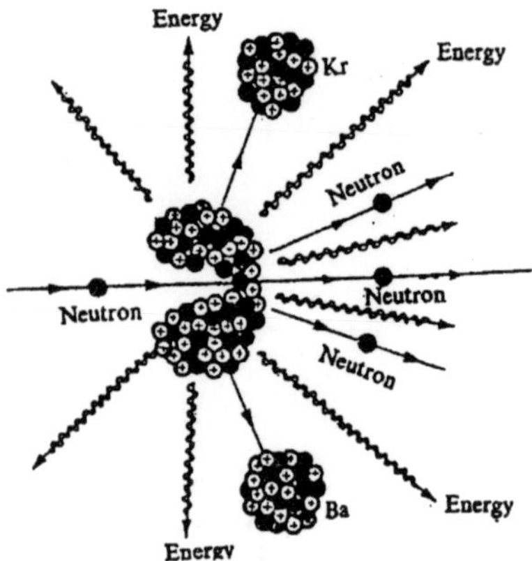

Thus when a slow neutron strikes ${}_{92}U^{235}$ nucleus, it is captured and a highly unstable nucleus ${}_{92}U^{236}$ is formed which at once breaks up into two fragments with the emission of two or three fast neutrons. A wide range of fission fragments is possible. One of the typical fission reactions is

$${}_{92}U^{235} + on^{1} \rightarrow ({}_{92}U^{236}) \rightarrow {}_{56}Ba^{144} + {}_{36}Kr^{89} + 3\ {}_{0}n^{1}.$$

In general, the fission fragments are found to be the radioactive isotopes of elements lying in the mass number range roughly from 70 to 160. All fragments undergo several β–decays (electron emissions) until they reach some stable end-product.

Later researches showed that besides uranium, other nuclei are also assignable. Thorium (${}_{90}Th^{232}$) and protactinium (${}_{91}Pa^{231}$) can be fissioned by fast neutrons, whereas the transuranium element plutonium (${}_{94}Pu^{239}$) and an artificial isotope of uranium (${}_{92}U^{238}$) can be fissioned by thermal neutrons.

Besides neutrons, accelerated protons, deuterons and α–particles can also induce fission in the nuclei of thorium, uranium and transuranium elements. Even γ-rays can cause nuclear fission which is known as 'photo-fission. Some lighter elements can also be fissioned by very high energy photons and deuterons.

'Spontaneous' fission (in which no bombarding particle is required) of thorium, uranium and transuranic elements has also been detected. It is a process (like natural radioactivity) in which a nucleus splits into two fragments of its own accord. The probability of this type of fission is, however, very small.

Bohr-Wheeler Theory of Nuclear Fission: We know that U^{238} can be fissioned by fast neutrons only (having energy 1.0 MeV or more), while U^{235} is fissionable by fast as well as by slow neutrons. An explanation of this observation was given by Bohr and Wheeler in terms of the liquid-drop model of the nucleus.

A nucleus is like a spherical drop of an incompressible liquid which is held in equilibrium by a balance between the short-range, attractive forces between the nucleons and the repulsive electrostatic forces between the protons. The internucleon forces also give rise to surface tension forces which maintain the "spherical" shape of the drop (Fig.).

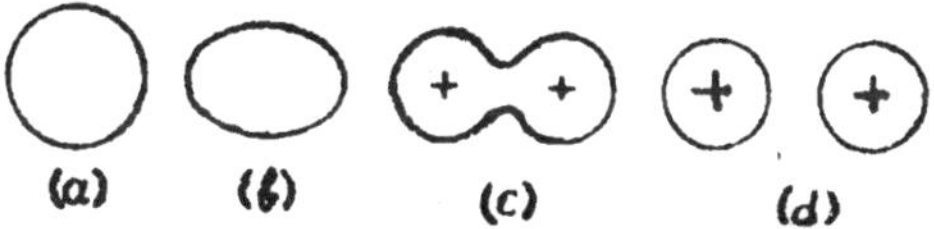

When the nucleus-drop captures a neutron, it becomes a compound nucleus of very high energy. The energy added to the nucleus is partly the kinetic energy of the incident neutron and partly the binding energy of the same neutron which it releases on being captured. This excitation energy initiate rapid oscillations within the drop which at times become ellipsoidal in shape (Fig. b). The surface tension forces tend to make the drop return to its spherical shape, while the excitation energy tends to distort the shape still further. If the excitation energy is small, the nucleus oscillates until it eventually loses its excitation energy by γ-emission (radiative capture) and returns to the spherical shape. If the excitation energy is sufficiently large, however, the nucleus may attain the shape of a dumbbell (Fig. c). When this happens, the repulsive electrostatic forces between the two parts of the dumbbell overcome the attractive forces between nucleons and the splitting takes place. Each splitted part becomes spherical in shape (Fig. d). Thus there is a threshold *activation* energy required to produce stage (c) after which the compound nucleus is bound to split.

The activation energy required to induce fission in various heavy nuclei can be calculated from the liquid-drop model. It gives following values for the Uranium nuclei:

Target Nucleus	Compound Nucleus	Activation Energy
${}_{92}U^{235}$	$[{}_{92}U^{236}]$	6.4 MeV
${}_{92}U^{238}$	$[{}_{92}U^{239}]$	6.6 MeV

When a neutron is captured by one of these nuclei, an energy equal to the corresponding activation energy must be supplied if fission is to occur. The actual energy supplied is equal to the sum of the kinetic energy and the binding energy of the neutron. For U^{235}, the binding energy released by the captured neutron in forming U^{236} is 6.8 MeV. This is *greater* than the required 6.4 MeV of energy. Hence fission would occur even if the kinetic energy of the incident neutron is near zero. This is why the thermal neutrons (energy 0.03 eV) cause fission in U^{235}.

For U^{238}, however, the binding energy released by the captured neutron in U^{239} is only 5.5 MeV which is *less* than the required 6.6 MeV by 1.1 MeV. The incident neutron must therefore have a kinetic energy of atleast 1.1 MeV to induce fission. This is the reason that U^{238} can be fissioned by fast neutrons only, with energy above 1 MeV. For the same reason the nuclei ${}_{92}U^{233}$ and ${}_{94}Pu^{239}$ are fissionable by slow neutrons, while ${}_{90}Th^{232}$ and ${}_{91}Pa^{231}$ can be fissioned by fast neutrons only.

Fission Fragments: In a given nucleus the fission may occur in a number of different ways. In general, the fissionable nucleus gives *two* fission fragments which thereafter decay by β–emission into stable end-products. What particular fragments are produced by the given nucleus is a matter of chance. For example, in case of the fission of U^{235} by thermal neutrons, a wide-range of primary fission-fragments having mass numbers roughly between 70 (Fig.) and 160 is possible. The mass distribution of the fission fragments may be shown by a 'fission yield curve', in which the percentage yields of the different fragments are plotted against mass number (Fig.). The curve shows that the splitting of U^{235} nucleus into two fragments of equal mass (A ≈118) has only a 0.01% chance of occurring; whereas the formation of fragments with mass numbers 95 and 140 is most likely (7%).

About 97% of the U^{235} nuclei undergoing fission give fragments which fall into two groups, a lighter group with mass numbers from 85 to 104, and a heavier group with mass numbers from 130 to 149.

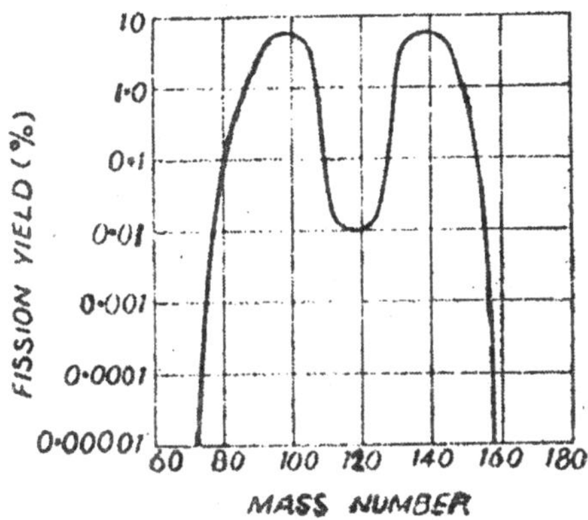

Neutron Emission in Nuclear Fission: An important feature of nuclear fission is the emission of fast neutrons. Most of them (more than 99%) are emitted almost instantaneously (≈ within 10^{-14} sec) with the fission process. These are called 'prompt'

neutrons. They have an energy distribution, with an average energy of 2 MeV. In addition, a few neutrons (less than 1% of the total) are emitted a short time later (ranging from 0.05 sec to 1 min) after the fission has occurred. These are called 'delayed' neutrons. On the average, 2.5 neutrons are emitted per fission.

The reason of the emission of the prompt neutrons is as follows. The heavy nuclei have a greater neutron/photon ratio than the medium-mass nuclei. Therefore when a heavy nucleus splits into lighter nuclei, the primary fission fragments have a n/p ratio which is too large for their stability. In other words, the fragments are overloaded with neutrons. Hence they liberate few neutrons as soon as they are formed. These are the prompt neutrons.

The liberation of prompt neutrons does not completely eliminate the neutron-overloading of fission fragments. Therefore as a further means of decreasing n/p ratio, the fragments undergo a chain of β–decays attended by the emission of γ-rays until stable end-products are reached. In a few cases neutrons are emitted

Energy Released in Nuclear Fission: The binding energy *per nucleon* for nuclei of intermediate mass is greater than that of heavy nuclei by about 1 MeV. Hence fission of a heavy nucleus into two lighter nuclei of intermediate masses is bound to release a large amount of energy. This energy release is the most striking aspect of nuclear fission.

A rough estimate of the energy released per fission can be made by using the binding energy curve (Fig.). The curve shows that heavy nuclei whose mass numbers are near 240 have binding energies of about 7.6 MeV per nucleon; while fission fragments whose mass numbers lie in the range roughly from 70 to 160 have binding energies of about 8.5 MeV per nucleon.

Thus the fission fragments have, on the average, binding

energy of about 0.9 MeV per nucleon greater than the (heavy) nucleus which has been fissioned. Hence 0.9 MeV/nucleon is released during the fission. The total amount of energy released per fission of a compound nucleus U^{236} (which has 236 nucleons) is thus

$$(0.9 \text{ MeV}) \times 236 = 212 \text{ MeV},$$

or roughly 200 MeV. This is much larger than the energy liberated in nuclear reactions (other than fission) which is of the order of 10 MeV. (The energy liberated in an ordinary chemical reaction is only a few electron volts. The energy released per fission is distributed roughly as follows :

Fission fragments (kinetic energy)	168 MeV
Prompt neutrons (kinetic energy)	5 MeV
Prompt γ-rays	5 MeV
β–decay energy from fission fragments	5 MeV
γ-rays from fission fragments	7 MeV
Neutrino energy from β–decay	10 MeV
	200 MeV

Thus we see that most of the (84%) energy released during fission goes into the kinetic energy of the fission fragments, 5% is given off as kinetic energy of neutrons and γ-rays which are emitted at the time of fission, and the remainder (11%) appears as radioactivity of the fission, fragments which decay to form stable end-products.

Chain-reaction: When a neutron fissions a uranium nucleus then, besides the fission fragments, a few fast neutrons are also emitted. If one or more of the emitted neutrons are used to fission other nuclei, further neutrons are produced and the process is repeated. The reaction thus becomes self-sustained and is known as a chain reaction (Fig.).

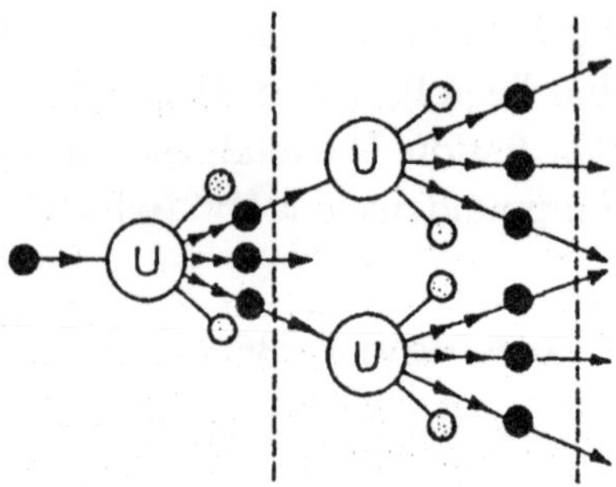

The chain reactions may be of two types:

(i) *Uncontrolled Chain Reaction:* If more than one of the neutrons emitted in a particular fission cause further fissions, then the number of fissions increases rapidly with time. Since a large amount of energy is liberated in each fission, within a very short time the energy takes a tremendous magnitude and is released as a violent explosion. This is what happens in a nuclear bomb. Such a chain reaction is called an "uncontrolled" reaction.

(ii) *Controlled Chain Reaction:* If by some means, the reaction is controlled in such a way that only one of the neutrons emitted in a fission causes another fission, then the fission rate remains constant and the energy is released steadily. Such a chain reaction is called a "controlled" reaction. It is used in a nuclear reactor.

The two types of chain reactions take place in different situations.

General Condition for a Self-Sustained Chain Reaction—Critical Size: In the fission of uranium nuclei, 2.5 neutron are, on the average, emitted per fission. Not all of these neutrons are available for further fissions. Some of them escape through the surface of the uranium, while many are lost in non-fission processes such as radiative capture by the nuclei of uranium and other materials present. Hence the basic condition that must be satisfied for sustaining a chain reaction is that *on the average, at least 'one' of the 2.5 neutrons born per fission must cause* *another fission.* This requirement may be stated by means of a constant k defined as:

$$k = \frac{\text{number of neutrons present in one generations}}{\text{number of neutrons present in the previous generation}}$$

k is known as 'neutron multiplication factor'. If $k < 1$, the chain reaction will slow down and stop. If k– 1, the reaction will proceed at a steady rate. If $k > 1$, however, the reaction will grow to an explosive rate. These situations are known as 'sub-critical' 'critical' and 'super-critical' respectively.

In a nuclear bomb, values of k considerably greater than 1 are needed; while in a nuclear reactor, k is required to be 1 or only slightly greater than 1.

In order to have the neutron multiplication factor greater than 1, the rate of production of neutrons must be larger than the rate of their loss. If the uranium is in the form of a solid sample, the rate of production of neutrons will be proportional to its volume while the rate of their escape will be proportional to the surface area. Since for a given volume, the sphere has the smallest surface area, the uranium should be taken in the form of a sphere.

Now, let us consider a uranium sphere of radius R. Let N_1 be the number of neutrons produced in a given time-interval, N_2 the number of neutrons lost in non-fission processes and N_3 the number escaped through the surface in the same time-interval. N_1 and N_2 will be proportional to the volume, while N_3 will be proportional to the surface area of the sphere. Thus

$$N_1 \propto \frac{4}{3}\pi R^3 = C_1 R^3,$$

$$N_2 \propto \frac{4}{3}\pi R^3 = C_2 R^3,$$

and $$N_3 \propto 4\pi R^2 = C_3 R^2,$$

where C_1, C_2 and C_3 are proportionality constants. For $K > 1$ we must have

$$N_1 > N_2 + N_3$$

or $$C_1 R^3 > C_2 R^3 + C_3 R^2$$

or $$C_1 R > C_2 R + C_3$$

or $$(C_1 - C_2) R > C_3$$

or $$R > \frac{C_3}{C_1 - C_2} = C \text{ (say)}.$$

C is known as the critical size of the sample. Thus, *in order to achieve a self-sustained chain reaction, the size (or mass) of the sample must be greater than a critical value* C. Below the critical size (or critical mass) the chain reaction would not occur.

Further, even if we take a piece of natural uranium in a size greater than the critical value, the chain reaction would not develop in it. The reason is that the neutrons emitted in a fission carry an energy of about 2 MeV. Such 'fast neutrons' have a much larger probability of being scattered by U^{238} than fissioning it. In this process their energy is reduced to below 1 MeV and then they are captured by U^{238} without fission. U^{235} nuclei, however, can be fissioned by fast neutrons but they are insignificantly small in number (only 0.7%) in natural uranium.

The isotope U^{235} can, however, be separated from natural uranium and then used as fissionable material. It can be fissioned by neutrons of *all* energies and has a much larger probability of fission than radiative capture. Hence, once a fission is initiated, the chain reaction is built up at an explosive rate.

Achievement of Controlled Chain Reaction: A different way of carrying out a chain reaction in 'natural' uranium is to rapidly slow down the emitted neutrons to thermal energies (≈ 0.03 eV) by means of moderators. The fission probability of U^{235} by thermal neutrons is very large. Therefore, inspite of the very small concentration of U^{235} in natural uranium, fissions of U^{235} occur slightly more frequently than the radiative (non-fission) capture of these neutrons both by U^{235} and U^{238}. Hence the neutron reproduction factor becomes very slightly greater than 1 and the chain reaction continues at almost a steady rate. This method is employed in a nuclear reactor.

Role of Moderator

A controlled chain-reaction is obtained in a nuclear reactor by slowing down the neutrons emitted in nuclear fissions. This is achieved by colliding the (fast) neutrons with such nuclei which have a small probability for neutron capture but a large probability for neutron scattering. A substance whose nuclei have these properties is called a 'moderator'. The nuclei of the moderator absorb energy from the colliding neutrons which are then scattered with reduced energy. The energy-loss is maximum if the nuclei are of the same mass as neutrons. Heavy water (D_2O) which has heavy-hydrogen nuclei, and pure graphite which has carbon nuclei are the best moderators. About 250 collisions are required to reduce the energy of a neutron from 2 MeV to thermal energy in heavy water and about 100 collisions in graphite. When heavy water is the moderator, the uranium is taken in the form of very small particles of uranium oxide suspended in the (heavy) water. The reactor using it is said to be 'homogeneous'. When graphite is the moderator, then uranium is taken in the form of rods distributed through the graphite in the form of a lattice. The reactor using it is said to be 'heterogeneous'.

Pu^{239} versus U^{235} as Fissionable Material: Pu^{239} has almost same fission properties as U^{235}. Hence both can be used for fissioning. But, however, the separation of U^{235} from natural uranium (99.3% U^{238} + 0.7% U^{235}) is not easy. The reason is that U^{235} and U^{238} have same chemical properties and so they cannot be separated from each other by any chemical method. They can be separated either by thermal diffusion or in a mass spectrograph. These processes are, however, very slow and expensive because of a very small fractional difference in the masses of U^{235} and U^{238} and the very small relative concentration of U^{235}.

Pu^{239} is produced by neutron bombardment of U^{238} and is a chemically different element. Therefore it can be easily separated from uranium by chemical methods. Hence Pu^{239} is now being used all over in nuclear bombs and nuclear reactors.

Power of Atom Bomb

An atom bomb, infact a nuclear fission bomb, is a device in which an *uncontrolled* chain reaction is built up in a fissionable material by means of fast neutrons. It then releases tremendous amount of energy in a very short time.

The bomb consists essentially of two pieces of U^{235} (or Pu^{239}), which are kept separated in a massive cover of a high-density material (Fig.). The mass of each piece is below the critical value so that any stay neutron, either from cosmic rays or produced by spontaneous fission, is unable to start a chain reaction. Thus, so long the two pieces are kept separated, they are perfectly stable and safe.

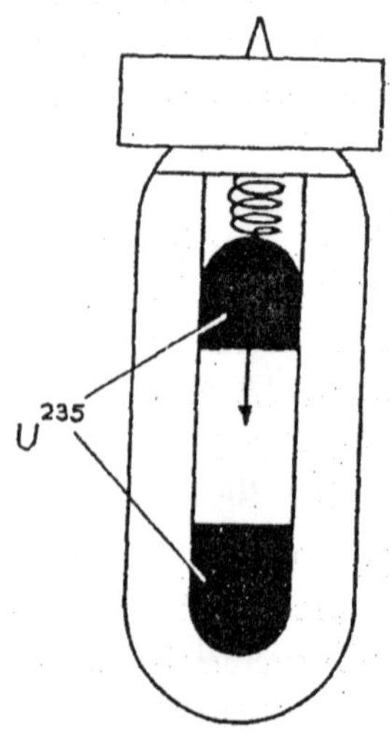

When an explosion is required, the two pieces are brought rapidly together (in a time less than a microsecond) by means of some device so that the total mass becomes supercritical *(greater than the critical value).* As soon as this happens, the stray neutrons initiate chain reaction in which fast neutrons rapidly multiply and within a few millionths of a second the reaction acquires an explosive nature.

An idea of the rate of energy released in a nuclear bomb having 1 kg of uranium can be made in the following way:

In a nuclear chain reaction the number of neutron N, available to produce fission at any time t is given by

$$N = N_0 e(k - 1)\ t/\tau$$

where k is the neutron multiplication factor and τ is the mean life-time of fission neutrons, τ is the time which elapses between the emission of a fission neutron and the production by this neutron of a further fission neutron. The average value of τ in a nuclear bomb is one nano-second (10^{-9} sec). Therefore the increase in the neutron population after 1 microsecond ($t = 10^{-6}$ sec) is given by (taking k =1.1).

$$\frac{N}{N_0} = e^{0.1} \times 10^{-6}/10^{-9} = e^{100} = 10^{43}.$$

Thus within a micro-second the neutrons multiply many millions of times which are sufficient *to* fission all the 10^{24} uranium nuclei present in 1 kg. Since each fission produces about 200 MeV of energy, an amount of 2×10^{26} MeV of energy could be released within 1 micro-second. This will constitute a large explosion.

In the explosion, a temperature of the order of 10^7 °C and a pressure of several million atmospheres is developed. Large quantities of radioactive material and blinding flashes of light are also produced. All objects and living creatures within a range of hundreds of kilometers are completely destroyed. The radioactive dust formed is carried away by air-currents to distant places causing loss beyond description. The bomb dropped on Hiroshima in 1945 had released an amount of energy equivalent to that from 20,000 tons of TNT. It was called a 20-kiloton bomb.

An upper limit is set to the size of a fission bomb by the practical difficulty of combining more than two pieces at exactly the same time, and because each of the pieces must be below the critical size so that it can be transported safely.

Functions of Nuclear Reactor

A nuclear reactor, or a nuclear pile, is a device in which a self-sustaining, *controlled* chain reaction is produced in a fissionable material. It is thus a source of controlled energy. The main parts of a modern reactor are as follows:

(i) *Fuel:* The fissionable material, known as fuel, plays the key role in the operation of a reactor. Uranium enriched with the isotope U^{235}, or Pu^{239}, is used as fuel.

(ii) *Moderator:* The moderator slows down the neutrons to thermal energies by elastic collisions between its nuclei and the fission neutrons. The thermal neutrons have a very high probability of producing fission in U^{335} nuclei. Heavy water, graphite and beryllium oxide are most suitable moderators.

(iii) *Control Rods:* These are the rods of cadmium (or boron) which are used to control the fission rate in the reactor. Cadmium and boron are good absorbers of slow neutrons. Therefore, when the rods are pushed into the reactor, the fission rate decreases, and when they are pulled out, the fission grows.

(iv) *Shield:* Since nuclear fissions produce various types of radiations which are dangerous to human life, the reactor is surrounded by a concrete wall about 2 meter thick and containing a high proportion of elements like iron.

(v) *Coolant:* The energy is released inside the reactor in the form of heat which is removed by means of a cooling agent, known as coolant. The carbon dioxide gas is the main coolant in a reactor. It is circulated through the interior of the reactor by a pumping system.

(vi) *Safety Device:* In an emergency, if the reactor begins to go too fast, a special set of control rods, known as shut-off rods, drop in automatically. They immediately absorb the neutrons so that the chain reaction stops entirely.

Construction: One type of nuclear reactor is shown in Fig. It consists of a large number of uranium rods placed in a calculated geometrical lattice between layers of pure graphite (moderator) blocks. To prevent oxidation of uranium and also to preserve the gaseous fission products, the rods are covered by close-fitting aluminium cylinders. The control rods are so inserted in the lattice that they can be raised or lowered between the uranium rods whenever necessary. The whole reactor is surrounded by a concrete shield.

Working: The actual operation of the reactor is started by raising the control rods so that they do not absorb many neutrons. Even a single neutron is capable of starting fission and there are always a few stray neutrons present either from the cosmic radiation or from a spontaneous fission. As soon as a neutron strikes a U^{235} nucleus and fissions it, two or three fast neutrons are emitted. These neutrons are slowed down from energies of several MeV to energies of less than 1 eV by collision with moderator nuclei, after which they induce further fission in U^{235}. The reaction, once started, is controlled by moving the control rods in and out.

Carbon dioxide is pumped rapidly through the reactor to carry away the heat generated by the fission of the uranium nuclei. The hot carbon dioxide gas passes through a heat-exchanger where it gives up its heat to water and converts it into steam. This steam drives the turbines and generates electric power.

The size of the reactor may be reduced by using heavy water as a moderator (in place of graphite). It, being lighter, slows the neutrons more effectively.

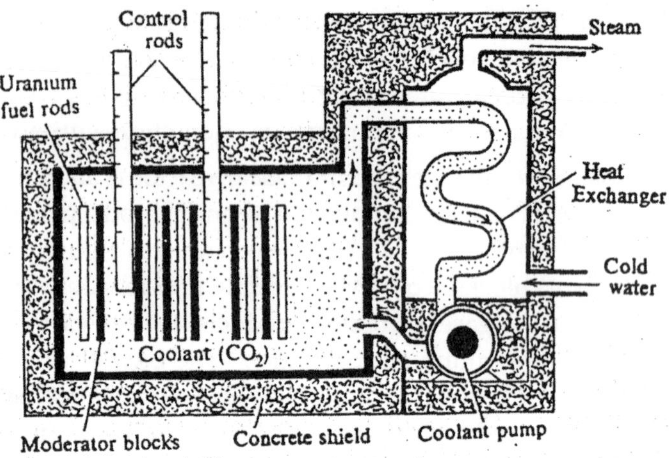

Neutron Cycle: The neutron balance in a thermal reactor can be described in terms of a cycle. For simplicity we assume that the reactor is infinitely large so that there is no leakage of neutrons through its surface. Let us start with the fission of U^{235} nucleus by a thermal neutron. In this process, suppose N_0 fast neutrons are emitted. These neutrons have an energy above the fission threshold for U^{238} and so some of these may produce fission in U^{238} before colliding with moderator nuclei, causing a fractional increase $\in$. The number of fast neutrons available is now $N_0\in$ where $\in$ is called the fast fission factor', $\in$ is usually about 1.03.

The $N_0\in$ neutrons diffuse through the pile and are slowed down by collisions with moderator nuclei. However, a few of them are captured by U^{288} before they are slowed down to thermal energies. If p is the fraction which escapes resonance capture, the number of thermal neutrons now available is $N_0\in p$, where p is called 'resonance probability' p is usually 0.95.

Of these $N_0 \in p$ thermal neutrons, only a fraction f may succeed in producing fission in U^{235}, others being lost by absorption in other materials, such as the moderator, the control rods, the metal casing, impurities in uranium, etc. Thus the number of U^{235} nuclei undergoing fission is $N_0 \in pf$, where f is called the 'thermal utilisation factor', and is always less than 1.

If each thermal fission in U^{235} produces η 'fast' neutrons to start the cycle over again, then the total number of neutrons after one cycle, or one generation, is

$$N = N_0 \in pf\eta.$$

The multiplication factor for a reactor (of infinite dimensions) is therefore

$$k = \frac{N}{N_0} = \in pf\eta.$$

This formula is called the 'four-factor formula.'

For the steady operation of the reactor at a particular power level, k must be equal to 1. The reactor is then said to be 'critical. The condition is achieved by varying the factor f by adjustment of control rods. If $k < 1$ then the chain reaction stops and the reactor becomes sub-critical. If $k > 1$, the fissions increase from cycle to cycle and the reactor becomes supercritical.

Neutrons and their Part

While most of the fission neutrons are emitted promptly, about 0.7% are emitted a short time later after the fission has occurred. These delayed neutrons help in increasing the over-all average life-time of fission neutrons. The average life-time of the prompt neutrons in a graphite reactor is about 1 milli-second (10^{-3} sec) but with the inclusion of the delayed neutrons the over-all average increases to 0.1 second. Thus sufficient time is available for adjusting the control rods. If *all* the neutrons were prompt, the reactor would become supercritical and would be unmanageable.

Applications: The nuclear reactors are used mainly for the following purposes:

(i) *Production of Pu^{239}:* The ordinary uranium reactor is used to produce plutonium (Pu^{239}) which is a better fissionable material than U^{235}. The neutrons in the reactor which do not participate in the fission chain-reaction of U^{235}, are absorbed by U^{238} and convert it into a heavier isotope U^{239}:

$$_{92}U^{238} + {}_0n^1 \rightarrow {}_{92}U^{239} + \gamma \text{ (energy)}$$

$_{92}U^{239}$ is an unstable nucleus. It emits a β–particle and is converted into a new heavier element, neptunium ($_{93}Na^{239}$):

$$_{92}U^{239} \rightarrow {}_{93}Np^{239} + {}_{-1}\beta^0 + \overline{v}$$
(anti-neutrino)

Neptunium also emits a β–particle and is converted into plutonium ($_{94}Pu^{239}$):

$$_{93}NP^{239} \rightarrow {}_{94}Pu^{239} + {}_{-1}\beta^0 + \overline{v}.$$

Since plutonium is a new element and is different in chemical properties from uranium, so it can easily be separated from uranium. It is used in nuclear reactors and in nuclear bombs in place of U^{235}. The critical size of Pu^{239} is smaller compared to that of U^{235} for the fission chain-reaction. Therefore, the use of Pu^{239} is economical compared to U^{235}.

(ii) *Production of Neutron Beam:* Fast-moving neutrons are emitted by the fission of U^{235} in the reactor. By converging these neutrons into a fine beam, artificial disintegration of other elements are studied.

(iii) *Production of Radio-Isotopes:* Artificial radioactive isotopes of many elements are produced in the reactor. For this, the element is placed in the reactor and bombarded with fast-moving neutrons. These radio-isotopes are utilised in medicines, biology, agriculture, industries and scientific discoveries.

(iv) *Generation of Energy:* The energy produced in reactors is used to run electric-generators. Thus nuclear energy is converted into electrical energy. It is used at power stations to generate electricity on a large scale which runs industries. Nuclear energy can be used in place of coal and patrol for driving the engines and for the propulsion of ships, submarines and aircrafts.

Thermal Reactor and Breeder Reactor: The reactors in which energy is produced by the fission of U^{235} by slow neutrons, are called 'thermal reactors'. Since the major part in ordinary uranium is U^{238} (U^{235} is only 0.7%), therefore the fission of U^{235} is very costly, and this would lead to an early depletion of uranium reserves.

We know that besides U^{235}, Pu^{239} and U^{238} are also fissionable substances. So Pu^{239} is produced from U^{238} in many nuclear reactors. In addition to it, U^{233} is also produced from Th^{232}. The quantity of the substance produced (Pu^{239} and U^{233}) in these reactors is more than the quantity of substances (U^{238} and Th^{232}) consumed. Such reactors are called breeder reactors'.

What is Fusion?

When two or more very light nuclei moving at very high speeds are fused together to form a single nucleus, then this process is known as 'nuclear fusion'. The mass of the product nucleus is less than the sum of the masses of the nuclei which are fused. The lost mass is converted into energy which is released in the process. This property of the light nuclei is shown by the binding energy curve (Fig.), in which the average binding energy per nucleon rises rapidly with increase in mass number in the range of low mass number nuclei.

For example, two deuterons (heavy-hydrogen nuclei) can

be fused to form a tritron (tritium nucleus) according to the following reaction:

$$_1H^2 + {_1H^2} \rightarrow {_1H^3} + {_1H^1} + 4.0 \text{ MeV energy.}$$

The tritron so formed can further fuse with a third deuteron to form an α–particle (helium nucleus):

$$_1H^3 + {_1H^2} \rightarrow {_2He^4} + {_0n^1} + 17.6 \text{ MeV energy.}$$

The net result of the two reactions is the burning of three deuterons and the formation of an α–particle ($_2He^4$), a neutron ($_0n^1$) and a proton ($_1H^1$). The total energy released is 21.6 MeV, so that the energy released per deuteron burnt is 7.2 MeV. Most of the liberated energy appears as kinetic energies of neutron and proton.

Alternatively, the fusion of three deuterons into an α–particle can take place as follows:

$$_1H^2 + {_1H^2} \rightarrow {_2He^3} + {_0n^1} + 3.3 \text{ MeV energy,}$$

$$_2He^3 + {_1H^2} \rightarrow {_2He^4} + {_1H^1} + 18.3 \text{ MeV energy.}$$

The net result is same as before; the energy released per deuteron burnt being again 7.2 MeV.

The energy output in a fusion reaction (21.6 MeV) is much less than the energy released in the fission of a U^{235} nucleus which is about 200 MeV, But this does not mean that fusion is a weaker energy-source than fission. The number of deuterons in 1 gram of heavy hydrogen is much larger than the number of U^{235} nuclei in 1 gm of uranium. Therefore, *the energy output per unit mass of the material consumed is much greater in case of the fusion of the light nuclei than in case of the fission of heavy nuclei.*

The fusion process is, however, not easy to carry out. Since the nuclei *to* be fused are positively charged, they would repel one another strongly. Hence they must be brought very close together not only by high pressure but also with high kinetic energies ($\approx$ 0.1 MeV). For this, a temperature of the order of 10^8 Kelvin is required. Such high temperatures are available in the sun and stars. On earth they may be produced by exploding a nuclear fission bomb. Since very high temperatures are needed for the fusion of nuclei, the process is called a 'thermonuclear reaction', and the energy released is called as 'thermonuclear energy'.

Difference between Nuclear Fission and Nuclear Fusion: These two nuclear processes are opposite in nature but give a common result namely, the liberation of large amount of energy. The fission is the splitting of a 'heavy' nucleus under neutron bombardment into lighter radioactive nuclei whose combined mass is less than the mass of the original nucleus. The lost mass is converted into energy. The energy when released in an uncontrolled manner produces destruction (as in a bomb), but if released steadily it can be used for peaceful purposes (as in a reactor).

The fusion, on the other hand, is the joining of two or more 'light' nuclei into a single nucleus whose mass is less than the sum of the masses of the joining nuclei. Again, the lost mass is converted into energy. This process happens at very high temperature and under very high pressure. The energy released in fusion is uncontrolled, though efforts are being made to control the fusion energy.

Importance of Hydrogen Bomb

It is a nuclear fusion bomb (also called as thermonuclear bomb), based upon the fusion of heavy-hydrogen nuclei. Since fusion takes place under the extreme conditions of high pressure and high temperature, a fission bomb must be used as igniter of a fusion bomb.

The central core of a hydrogen bomb is a uranium (or plutonium) fission bomb which is surrounded by the fusable material such as lithium hydride (LiH^2), a compound of heavy hydrogen. Thus, in turn, is surrounded by uranium. When the central fission bomb explodes, enormous temperature and pressure are produced and the fusion of surrounding heavy-hydrogen nuclei starts. The fast neutrons produced in fusion produce further fissions in the outer layer of uranium. Thus a fusion chain reaction is developed and a tremendous amount of energy is released.

The size of a fission bomb cannot be increased beyond a limit because in it the fissionable material is kept in two pieces, and the size of each piece should be less than the critical size. There is no such restriction on the size of hydrogen bomb. The material to be fused may be taken in it in any quantity. Once the fusion is initiated, it can spread throughout any mass of the material. Therefore, a fusion bomb is much more destructive than the fission bomb.

The fusion of nuclei is an uncontrolled process. Till now, there is no available method of controlling the release of fusion energy. Therefore, its only use so far is destructive.

Practical Difficulties in Controlling Thermonuclear (Fusion) Energy: A controlled release of fission energy is possible in a nuclear fission reactor. But there is not yet available any method of controlling the release of fusion energy. Hence a nuclear fusion reactor is so far only a dream.

To carry out controlled nuclear fusion, it is necessary to set up and maintain a temperature of the order of 10^8 K in a limited volume. At such a high temperature the atoms are entirely stripped of their electrons so that the fusable material is a completely ionised gas called 'plasma. The plasma thus consists of atomic nuclei and electrons in rapid random motion. The main problem is the design of a "container" in which the hot plasma can be contained under the required high pressure. No vessel of any substance can be a container because it would immediately evaporate.

Moreover, the contact of plasma with the walls of the vessel will result in its cooling. Attempts are being made to retain the plasma is a circular magnetic field created by means of a current pulse of the order of a million amperes (Fig.). The magnetic forces on the moving charged plasma particles make them travel along paths in a limited part of space. This is known as "pinch effect."

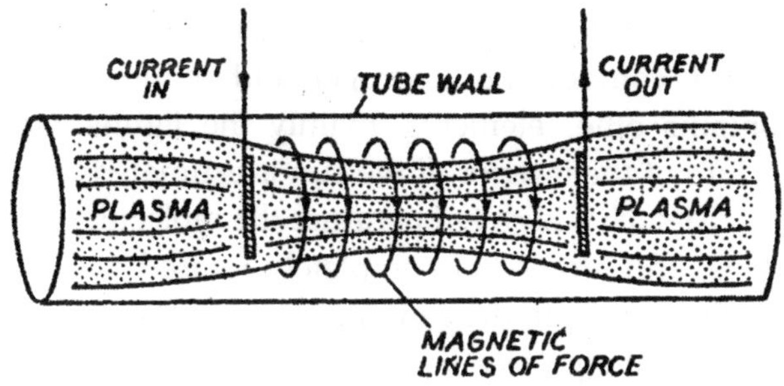

The achievement of controlled fusion will provide mankind with almost limitless source of energy. This is because the main fusion fuel is heavy hydrogen which is found in abundance in the sea water (whereas the fission fuel uranium is limited). Further, the products of fusion are, in general, not radioactive. Therefore, the problem of the disposal of the radioactive waste will not arise.

Benefits of Solar Energy

The sun has been radiating energy at an enormous rate for billions of years. The source of the continuous supply of energy remained a mystery for a long time. Chemical reactions in the sun cannot supply this energy. The reason is that even if the sun were composed entirely of carbon, then its complete combustion would supply energy at the present rate for only a few thousand years. In that case the sun would have burnt long ago. Helmholtz suggested that the sun is continuously contracting and so its gravitational energy is being converted into heat energy. But calculations show that contraction could supply not more than one percent of the actual energy output.

Hence gravitational contraction can also not be a possible source of solar energy.

Nuclear fission of heavy nuclei must also be ruled out because the abundance of heavy elements in the sun is too small to account for the observed rate of energy emission.

Infact, the process responsible for the solar energy is the fusion of light nuclei into heavier nuclei. About 90% of the solar mass is composed of hydrogen and helium, and the rest 10% contains other elements, mainly the lighter ones. The temperature of the interior of the sun is estimated to be about 2×10^7 K. At such a high temperature the molecules dissociate into atoms, and atoms are completely ionised to form a hot plasma. Fusion of hydrogen nuclei into helium nuclei is continuously taking place in this plasma, with the continuous liberation of energy. It is unlikely in the solar conditions that four hydrogen nuclei would fuse together directly to form a helium nucleus. This may take place through a cycle of processes, however. Two such cycles have been proposed; the carbon cycle and the proton-proton cycle.

Carbon Cycle: This cycle was proposed by Bethe in 1939 to account for the energy radiated by the sun. In this cycle the fusion of hydrogen nuclei into helium nucleus takes place in the sun through a series of nuclear reactions in which carbon acts as a catalyst. These reactions occur in the following order:

$$_6C^{12} + {}_1H^1 \rightarrow {}_7N^{13} + \gamma \text{ (energy)}$$

$$_7N^{13} \rightarrow {}_6C^{13} + {}_{+1}\beta^0 + v \text{ (neutrino)}$$

$$_6C^{13} + {}_1H^1 \rightarrow {}_7N^{14} + \gamma$$

$$_7N^{14} + {}_1H^1 \rightarrow {}_8O^{15} + \gamma$$

$$_8O^{15} \rightarrow {}_7N^{15} + {}_{+1}\beta^0 + v$$

$$_7N^{15} + {}_1H^1 \rightarrow {}_6C^{12} + {}_2He^4$$

$$\text{on adding:} \quad 4{}_1H^1 \rightarrow {}_2He^4 + 2\,{}_{+1}\beta^0 + 2v + \gamma \text{ (energy)}$$

The net result is the fusion of four hydrogen nuclei into a helium nucleus with the emission of two positrons ($_{+1}\beta^0$) and 24.7 MeV of energy. The nucleus $_6C^{12}$ which starts the cycle reappears in the final reaction. The emitted positrons combine with two electrons and are annihilated, producing about 2 MeV of energy. Thus about 26.7 MeV of energy is released for every helium nucleus formed. Enormous number of such fusions can simultaneously take place in the sun.

Proton-Proton Cycle: Another possible cycle of reactions is the proton-proton cycle which occur in the following sequence:

$$_1H^1 + {}_1H^1 \rightarrow {}_1H^2 + {}_{+1}\beta^0 + v \text{ (neutrino)} \quad \ldots (i)$$

$$_1H^2 + {}_1H^1 \rightarrow {}_2He^3 + \gamma \text{ (energy)} \quad \ldots (ii)$$

$$_2He^2 + {}_2He^3 \rightarrow {}_2He^4 + {}_1H^1 + {}_1H^1 \quad \ldots (iii)$$

Multiplying each of the equations and (ii) by 2, and then adding all the three, we get

$$4\ {}_1H^1 \rightarrow {}_2He^4 + 2_{+1}\beta^0 + 2v + \gamma \text{ (energy)}$$

In this cycle of reactions, the reactions (*i*) and (*ii*) must occur twice for the reaction (*iii*) to take place. The net result is again the fusion of four hydrogen nuclei into a helium nucleus with the emission of two positrons. Approximately same energy is released per cycle as in the carbon cycle.

In the sun, whose interior temperature is estimated to be 2×10^7 K; the proton-proton cycle has the greater probability for occurrence. In general, the carbon cycle is more efficient at high temperatures, while the proton-proton cycle is more efficient at low temperatures. Hence stars hotter than the sun obtain their energy (known as stellar energy) largely from the carbon cycle, while those cooler than the sun obtain the greater part of their energy from the proton-proton cycle.

The sun is emitting energy at the rate of about 4×10^{26} joule/second. Because of this, the solar mass is reducing at the rate of about 4×10^9 kg/second. Thus, the sun is annihilating at a very high rate. But the annihilated mass is still very much less than the total mass of the sun which is about 2×10^{30} kg. It is estimated that the sun will be emitting energy at the present rate for the next one thousand crore (10^{11}) years.

PROBLEMS

1. *In a nuclear reactor, fission is produced in 1 gm of U^{235} (235.0439 amu) in 24 hours by a slow neutron (1.0087 amu). Assuming that $_{36}Kr^{92}$ (91.8973 amu) and $_{56}Ba^{141}$ (140.9139 amu) are produced in all reactions and no energy is lost, write the complete reaction and calculate the total energy produced in MeV and in killowatt-hour. Given 1 amu = 931.5 MeV.*

Solution: The nuclear fission reaction is

$$_{92}U^{235} + {}_0n^1 \rightarrow {}_{56}Ba^{141} + {}_{36}Kr^{92} + 3\ {}_0n^1.$$

The sum of the masses before reaction is

$$235.0439 + 1.0087 = 236.0526 \text{ amu.}$$

The sum of the masses after the reaction is

$$140.9139 + 91.8973 + 3\,(1.0087) = 235.8373 \text{ amu.}$$

The mass loss in the fission is

$$\Delta m = 236.0526 - 235.8373 = 0.2153 \text{ amu.}$$

The energy equivalent of 1 amu is 931.5 MeV. Therefore, the energy released in the fission of a U^{235} nucleus

$$= 0.2153 \times 931.5 = 200 \text{ MeV.}$$

The number of atoms in 235 gm of U^{235} is 6.02×10^{23} (Avogadro number). Therefore, the number of atoms in 1 gm

$$= \frac{6.02 \times 10^{23}}{235} = 2.56 \times 10^{21}.$$

Hence the energy released in the fission of 1 gm of U^{235} *i.e,* in 2.56×10^{21} fissions is

$$E = 200 \times 2.56 \times 10^{21} = 5.12 \times 10^{23} \text{ MeV.}$$

Now, $$1 \text{ MeV} = 1.6 \times 10^{-13} \text{ joule.}$$

$$\therefore \quad E = 5.12 \times 10^{23} \times (1.6 \times 10^{-13})$$

$$= 8.2 \times 10^{10} \text{ joule.}$$

Now, 1 kilowatt-hour= 1000 watt x 3600 sec

$$= 1000 \frac{\text{joule}}{\text{sec}} \times 3600\text{sec}$$

$$= 3.6 \times 10^{6} \text{ joules.}$$

$$\therefore \quad E = \frac{8.2 \times 10^{10}}{3.6 \times 10^{6}} = 2.28 \times 10^{4} \text{ kW-H.}$$

2. *Calculate the energy released from the fission of 100 gm of U^{235} if the fission of one U^{235} nucleus gives 200 MeV of energy.*

[Ans: 5.12×10^{25} Mev]

3. *If in a certain fission process, the mass loss is 0.1%, then calculate the energy liberated by the fission of 1 kg of the substance. How much kilowatt-hour electric energy can be generated from it ?*

Solution: The mass loss in 1 kg of the substance is

$$\Delta m = \frac{0.1}{100} \times 1 \text{ kg} = 0.001 \text{ kg.}$$

According to Einstein's mass-energy relation, the energy liberated is

$$\Delta E = (\Delta m)\, c^2$$

$$= 0.001 \text{ kg} \times (3.0 \times 10^{8} \text{ m/s})^2$$

$$= 9.0 \times 10^{13} \text{ joule.}$$

Now, 1 kilowatt-hour = 3.6×10^6 joules.

$$\therefore \qquad \Delta E = \frac{9.0 \times 10^{13}}{3.6 \times 10^6} = 2.5 \times 10^7 \text{ kilowatt-hour.}$$

4. ***Calculate the useful power produced by a reactor of 40% efficiency in which 10^{14} fissions are occurring each second and the energy per fission is 250 MeV. Take 1 MeV = 1.6×10^{-13} joule.***

[Ans. 1.6 kilowatts]

5. ***Calculate the fission rate of U^{235} required to produce 2 watts and the amount of energy that is released in the complete fission of 0.5 kg of U^{235}. The energy released per fission of U^{235} is 200 MeV.***

Solution: The power to be produced is

$$P = 2 \text{ watts} = 2 \text{ joule/sec}$$

$$= \frac{2}{1.6 \times 10^{-13}} \text{ MeV/sec}$$

[$\because$ 1 MeV = 1.6×10^{-13} joule]

$$= 1.25 \times 10^{13} \text{ MeV/sec.}$$

The energy released per fission is 200 MeV.

$\therefore$ required no. of fissions per sec

$$= \frac{1.25 \times 10^{13}}{200} = 6.25 \times 10^{10}$$

The fission rate is 6.25×10^{10} per sec.

The number of atoms in 235 gm of U^{235} is 6.02×10^{23}. Therefore, the number of atoms in 0.5 kg (=500 gm) of U^{235}

$$= \frac{6.02 \times 10^{23}}{235} \times 500 = 1.28 \times 10^{24}.$$

The energy released per fission is 200 MeV. Therefore, the total energy released in the complete fission of 0.5 kg of U^{235}

$$= 200 \text{ MeV} \times (1.28 \times 10^{24})$$

$$= 2.56 \times 10^{26} \text{ MeV}.$$

6. ***Energy released in the fission of a single uranium nucleus is 200 MeV. Calculate the number of fissions per second to produce 1 milliwatt power.***

[Ans. 3.125 ×10⁷]

7. ***A reactor is developing nuclear energy at a rate of 32,000 kilowatts. How many atoms of U^{235} undergo fission per second? How many kg of U^{235} would be used up in 1000 hours of operation ? Assume an average energy of 200 MeV released per fission. Take Avogadro's number as 6×10^{13} and 1 MeV = 1.6×10^{-23} joule.***

Solution: The power developed by the reactor is 32000 kilowatts, i.e. 3.2×10^7 watts. Therefore, the energy released by the reactor per second is

$$= 3.2 \times 10^7 \text{ joules} \quad [\because 1 \text{ watts} = 1 \text{ joule/sec}]$$

$$= \frac{3.2 \times 10^7}{1.6 \times 10^{-13}} \quad [\because 1.6 \times 10^{-13} \text{ joule} = 1 \text{ MeV}]$$

$$= 2.0 \times 10^{20} \text{ MeV}.$$

The energy released per fission is 200 MeV. Therefore, the number of fissions occurring in the reactor per second

$$= \frac{2.0 \times 10^{20}}{200} = 1.0 \times 10^{18}.$$

The number of atoms (or nuclei) of U^{235} consumed in 1000 hours

$$= 1.0 \times 10^{18} \times (1000 \times 3600)$$

$$= 36 \times 10^{23}.$$

Now, 1 gm-atom (*i.e.* 235 gm) of U^{235} has 6×10^{23} atoms. Therefore, the mass of U^{235} consumed in 1000 hours is

$$= \frac{36 \times 10^{23}}{6 \times 10^{23}} \times 235$$

$$= 1410 \text{ gm} = 1.41 \text{ kg}.$$

8. Calculate the approximate mass of uranium which must undergo fission to produce same energy as is produced by the combustion of 10^5 kg of coal. Heat of combustion of coal is 8000 kcal/kg; the energy released per fission of U^{235} is 200 MeV and Avogadro's number is 6.02×10^{23} per gm-atom (1 cal = 4.2 joules).

Solution: The energy produced by 10^5 kg of coal

$$= 10^5 \times 8000 \text{ kcal}$$

$$= 8 \times 10^{11} \text{ cal}$$

$$= 8 \times 10^{11} \times 4.2 = 3.36 \times 10^{12} \text{joule}$$

[$\because$ 1 cal = 4.2 joule

$$= \frac{3.36 \times 10^{12}}{1.6 \times 10^{-13}} = 2.1 \times 10^{25} \text{ MeV}$$

[$\therefore$ 1 MeV = 1.6×10^{-13} joule]

The energy released per fission is 200 MeV. Therefore, the number of fissions required for 2.1×10^{25} MeV energy is

$$\frac{2.1 \times 10^{25}}{200} = 1.05 \times 10^{23}.$$

Since 235 gm of U^{235} contains 6.02×10^{23} atoms, the mass of uranium containing 1.05×10^{23} atoms is

$$\frac{1.05 \times 10^{23}}{6.02 \times 10^{23}} \times 235 = 41 \text{ gm}.$$

9. *Certain stars obtain part of their energy by the fusion of three α–particles to form a $_6C^{12}$ nucleus. How much energy does each such reaction evolve ? The mass of helium atom is 4.00260 amu while the mass of an electron is 0.00055 amu. The mass of $_6C^{12}$ atom is 12.0000... amu by definition. (1 amu = 931.5 MeV)*

Solution: The mass of $_2He^4$ atom is 4.00260 amu and it has 2 electrons. Therefore, the mass of its nucleus (α-particle)

= 4.00260—mass of 2 electrons

= 4.00260 – (2 × 0.00055)

= 4.00260 – 0.00110

= 4.00150 amu.

The mass of $_6C^{12}$ atom is 12.00000 amu and it has 6 electrons. Therefore, the mass of $_6C^{12}$ nucleus

= 12.00000 – (6 × 0.00055)

= 12.00000 – 0.00330

= 11.99670 amu.

When 3 α-particles fuse in a $_6C^{12}$ nucleus, the mass-loss is

(3 × 4.00150) – 11.99670 = 0.00780 amu.

The equivalent energy is

0.00780 × 931.5 = 7.26 MeV.

6

Photoelectric Effect

Two Effects

Impact of Photoelectricity: When a beam of light of frequency in the blue or ultraviolet region falls on a metal plate, blow-moving electrons are emitted from the metal surface. This phenomenon is known as 'photoelectric effect' and the electrons emitted are known as 'photoelectrons'. The photoelectrons can be obtained from non-metals and gases also, provided that light of the appropriate frequency is used.

Experimental Observation: Lenard in 1900 studied the photoelectric effect experimentally. His apparatus (Fig.) consists of an evacuated tube having two plates, C and A. A varying potential difference can be applied between C and A by means of a potential divider and a commutator K. The p.d. and the current in the circuit can be read by a voltmeter and an ammeter respectively.

Dependence Upon Intensity: Light of an appropriate *fixed* frequency is allowed to fall on the plate C. Photoelectrons are emitted from the surface of C. When the potential V of the plate A is made positive relative to C, a steady saturation current flows in the circuit, whatever the (positive) value of *V*. This means that all the photoelectrons emitted from C are reaching *A*. If the intensity of the incident light is increased, the saturation current is found to increase in the same ratio. This shows that the number of photoelectrons emitted per second varies directly as the intensity of the incident light.

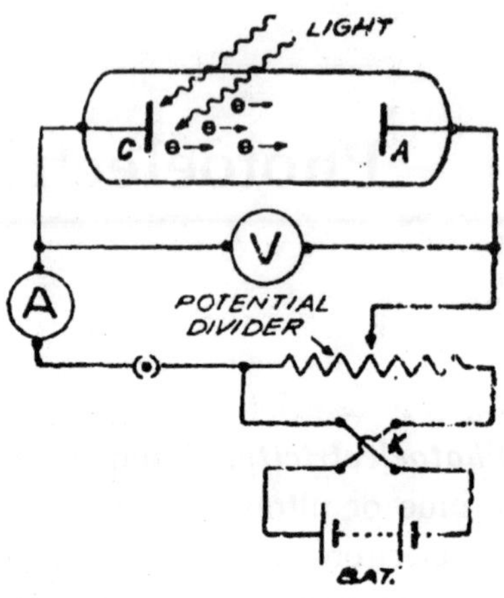

When the potential *V* of the plate *A* is made negative with respect to *C*, the electrons are retarted. If the potential is small, only the slower electrons are pulled back and the current falls. As the negative potential is increased, more and more faster electrons are pulled back. Ultimately, at a certain potential V_0, even the fastest electrons fail to reach *A*. The photoelectric current is then zero. An increase, however large, in the intensity of the incident light does not produce any current at the potential V_0 (Fig.). The (negative) potential V_0 *at* which the current is just reduced to zero is called the 'stopping potential. Since the energy of the fastest electrons is just annuled when they fall through the stopping potential, the stopping potential is a measure of the maximum kinetic energy of the photoelectrons. Hence we conclude that *the maximum energy of the photoelectrons is independent of the intensity of the incident light.*

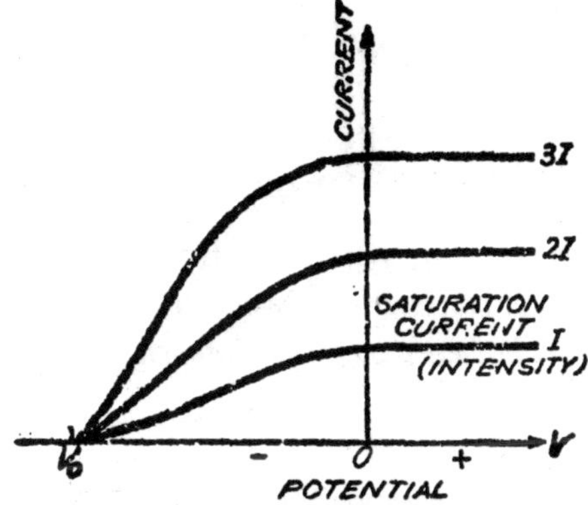

Dependence Upon Frequency: If, after reaching at the stopping potential, the incident light is replaced by another light of higher frequency, once more a current begins to flow in the circuit. This current again stops when the stopping potential is increased. Thus higher the frequency of the incident light, the higher is the stopping potential (Fig.) *i.e.* higher the maximum energy of the photo electrons.

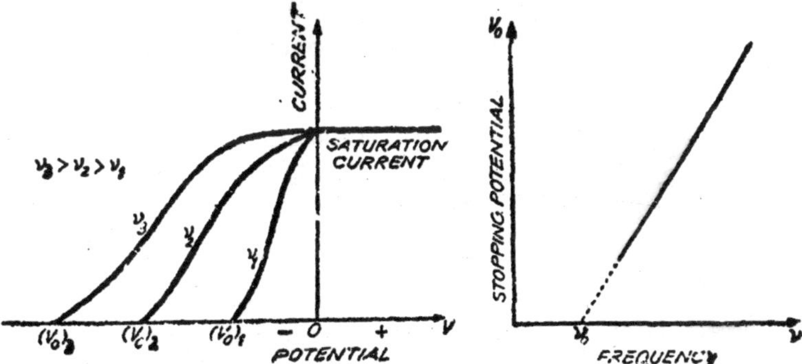

When a graph is plotted between the frequency v of incident light and the stopping potential $V_{0,}$ a straight line is obtained (Fig.). From this we conclude that *the maximum energy of the photoelectrons increase; linearly as the frequency of the incident tight increases.*

If the straight-line graph (between V_0 and v) is produced back, it cuts the v-axis at a point corresponding to a particular frequency v_0. This shows that *the emission of photoelectrons cannot occur if the frequency of the incident light is below a certain value* v_0, *however strong the intensity of light is* v_0 is called the 'threshold frequency and is different for different emitting surfaces.

On the other hand, when the frequency of the incident light is above v_0, *photoelectrons are emitted almost instantaneously with no time-lag, however pour the intensity of light.*

Colapse of Wave Theory

(*i*) According to the wave theory, the energy carried by a light beam is distributed uniformly and continuously over a wavefront *and is measured by the intensity of the beam.* Thus when light fails on a metal surface the energy of the wave should be transferred uniformly to the electrons in the surface before they are emitted out. Obviously, the energy taken up by the electrons must increase as the intensity of light is increased. This is against the experimental observation that the maximum energy of the emitted electrons is independent of the light intensity.

(*ii*) Again, the wave theory suggests that light of any frequency, however low it is, should be capable of ejecting electrons from a surface, provided only that the light is intense enough. Experiment, however, shows that light of frequency lower than a certain threshold value cannot eject photo electrons, no matter how intense it is.

(*iii*) Finally, the wave theory suggests that if the incident light is very feeble, the electron should take *appreciable* time before it acquires sufficient energy to come out from the surface. However, no detectable time-lag has ever been found between the falling of the light on the surface and the emission of the photo electrons. Thus wave theory fails in explaining the experimental observations regarding the photoelectric effect.

Quantum Theory: an Explanation

According to the quantum theory, the energy carried by a light beam of frequency v is concentrated in indivisible packets called 'photons', each photon carrying an amount *hv*. The intensity of the beam depends merely upon the number of photons. Now when light falls on a metal surface, a photon (energy *hv)* is *completely* absorbed by a single electron in the surface. Part of this energy is used in ejecting the electron against the attraction of the rest of the metal and rest is given to the electron as kinetic energy. If the electron doss not lose energy by internal collisions, it emits out, with a maximum kinetic energy. If the intensity of light is doubled, the number of photons falling per second on the surface will be doubled, the photon energy *hv* will remain the same) and hence the number of electrons emitted per second will and be doubled, their maximum energy remaining the same as before. Thus the number of photoelectrons emitted per second varies directly as the light intensity. But their maximum energy is independent of the intensity. This is completely is agreement with experiment

The maximum kinetic energy of the photoelectrons increases as the frequency of light, and hence the photon energy increases as observed by experiment.

The existence of a threshold frequency also speaks is favour of the photon (quantum) theory. The electron cannot leave the metal until the energy given to it by a single photon (*hv*) exceeds the energy required in liberating the electron against the attraction of the rest of the metal; no matter how many photons there are (that is no matter how intense the incident light is).

Finally, the absence of a time-lag also follows from the photon theory. As soon as the first photon strike the surface, in is immediately absorbed by *some* electron which escapes instantaneously, there being no time-lag.

The successful explanation of photoelectric effect in terms of quantum theory is one of the strongest evidences in favour of this theory.

Rules of Photoelectricity

There are following laws of photoelectric emission which have been derived from experimental observations:

(1) The number of photoelectrons emitted per second varies directly as the intensity of the incident light.

(2) The maximum energy of the photoelectrons is independent of the intensity of the incident light.

(3) The maximum energy of the photoelectrons increases linearly as the frequency of the incident light increases.

(4) The emission of photo electrons cannot occur if the frequency of the incident light is below a certain value v_o. (called the threshold frequency), however strong the intensity of light, v_0 is different for different emitting surfaces.

(5) Photoelectrons are emitted almost instantaneously, however poor the intensity of the incident light is.

Einstein's Photon Theory and the Photoelectric Equation: Einstein, in 1905, explained the photoelectric effect by introducing the photon theory of light. He assumed that the light (radiation) travels through pace in *indivisible* packets of energy called photons' or 'quanta', the energy of a single photon being *hv*, where *h* is the Planck's constant' and v the frequency of the light.

When light falls on a metal surface, a photon (energy *hv*) is completely absorbed by a single electron in the surface. Part of this energy is used in electing the electron against the attraction of the rest of the metal and the rest is given to the electron as kinetic energy. Those electrons which are ejected from some depth below the surface expend some of their acquired energy in collisions with the atoms. The energy lost in this way is a variable quantity and so the photo electrons are emitted with a range of kinetic energies.

Let us consider those photoelectrons which are emitted from very near the surface and do not suffer and collision. They emit with maximum kinetic energy $(\frac{1}{2}mv^2)_{max}$. Let W be the energy required to eject an electron against the attraction of the rest of the metal. W is called the work function' of the given metal. Thus we can write

$$hv = W + \left(\frac{1}{2}mv^2\right)_{max} \qquad \text{... (i)}$$

hv being the photon energy absorbed by the electron. Now, an electron cannot emit from the surface if the energy of the photon absorbed by it is less than W. If v_0 be the threshold frequency, then the threshold energy of the photon will be hv_0. Thus

$$W = hv_o$$

Making this substitution in eq. (*i*), we get

$$hv = hv_0 + \left(\frac{1}{2}mv^2\right)_{max}$$

or
$$\left(\frac{1}{2}mv^2\right)_{max} = h(v - v_0). \qquad \text{... (ii)}$$

This is Einstein's photoelectric equation.

Photoelectric Laws: Elaboration

If the intensity of light of a given frequency v is increased, the number of photons striking the surface per second will increase, but the energy hv of each photon will remain the same. Hence the number of electrons emitted per second will correspondingly increase, but their maximum, energy $(\frac{1}{2}mv^2)_{max}$ will remain the same, as is clear from Einstein's equation. Thus the laws (1) and (2) are explained.

It is also seen from Einstein's equation that the maximum energy of the photoelectrons, $(\frac{1}{2}mv^2)_{max}$, will increase linearity with the increase in the frequency v of the incident light. This is the photoelectric law (3).

The eq. (*ii*) further shows that if $v<v_0$, the kinetic energy of the photo electrons would be negative which is impossible. This means that photo electrons cannot emit if the frequency v of the incident light is less than the threshold frequency v_0. This is law (4).

Finally, as soon as the first photon strikes the surface, it is immediately absorbed by some electron which escapes instantaneously, there being no time-lag. This is law (5).

In Einstein's theory the emission of one electron is associated with the absorption of one photon. This, however, does not mean that every photon incident on the metal causes emission of an electron even though the photon energy is above threshold. The photon can be involved in many alternative processes, and hence the ratio of the number of electrons emitted to the number of photons incident on the surface is very much less than unity.

Millikan's Verification: Einstein's equation was tested by R. A. Millikan who measured the maximum energies of the photoelectrons emitted by a number of alkali metals over a wide range of light frequencies. His apparatus is shown in Fig.

Three different alkali metal surfaces are mounted on a drum *D* placed inside an evacuated chamber. The drum can be rotated from outside so that any surface can be scraped clean by a sharp knife *K* and then brought before a window *W**. A beam of monochromatic light passing through the window falls on the surface C. The emitted photo electrons are collected by an electrode *A* by means of a potential difference applied between C and *A* through a potential-divider. The galvanometer *G* measures this photoelectric current.

When the electrode A is given a negative potential with respect to C, the electrons emitted from C are retarted. If the potential is small, only the slower electrons are pulled back and the current falls. As the negative potential is increased, more and more faster electrons are pulled back. Ultimately, at a certain potential V_0, even the fastest electrons fail to reach A. The photoelectric current is then zero. The (negative) potential V_0 at which the current is just reduced to zero is called the 'stopping potential. Since the energy of the fastest electrons is just annuled when they fall through the stopping potential, we have

$$(\tfrac{1}{2}mv^2)_{max} = eV_0, \qquad \ldots (i)$$

where e is the electronic charge and V_0 is the stopping potential. Thus the stopping potential multiplied by electronic charge gives the maximum kinetic energy of the photoelectrons.

Millikan plotted a graph between the stopping potential and the frequency of light over a wide range of frequencies, and obtained a straight line (Fig.). Parallel lines were obtained for other metallic surfaces.

Now, according to Einstein's equation, we have

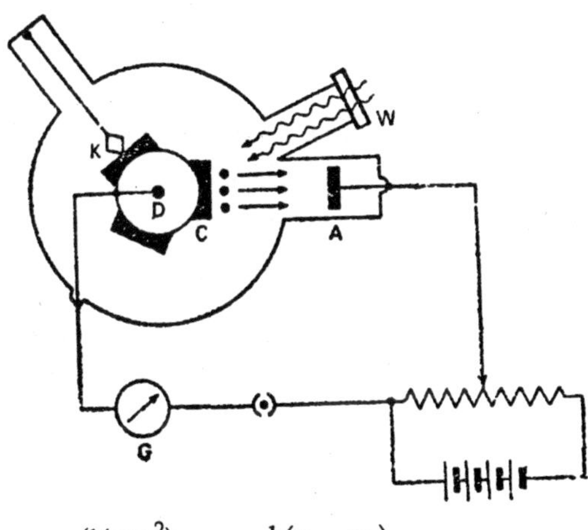

$$(\tfrac{1}{2}mv^2)_{\text{max}} = h(v - v_0).$$

But from (*i*), $(\frac{1}{2}mv^2)_{max} = eV_0$

$$\therefore \quad eV_0 = h(v - v_0)$$

or

$$V_0 = \left(\frac{h}{e}\right)v - \left(\frac{h}{e}\right)v_0 \ldots.$$

Since h and e are constants and v_0 is constant for a given surface, eq. (*ii*) indicates that the graph between the stopping potential V_0 and the frequency of light v must be a straight line. This is actually the case, as found by Millikan. Hence Einstein's equation is of the correct form.

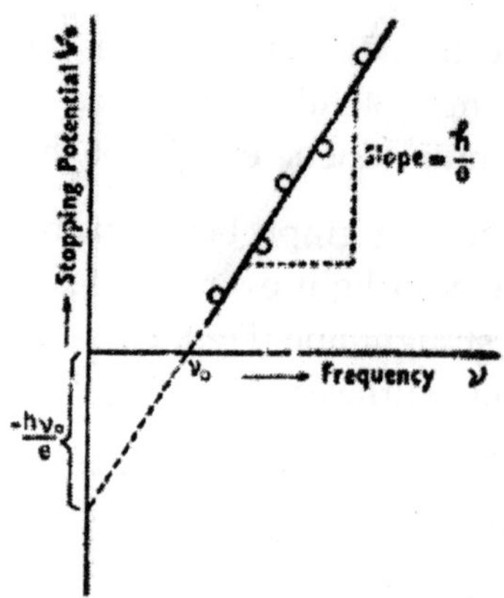

Determination of Planck's Constant (h): Eq. (*ii*) shows that the slope of the straight-line graph is h/e, and its intercept on the potential axis is $-hv_0/e$. Hence by measuring the slope of the straight-line obtained and, using the known value of e, the value of h can be obtained.

Determination of Threshold Frequency and Work Function:

Eq. (*ii*) shows that when $V_0 = 0; v = v_0$ That is, the intercept of the straight line on the v-axis gives the threshold frequency v_0 for the particular metal surface. This, when multiplied by h, gives the work function hv_0.

Let's see as to what is the effect on the energy and the number of emitted photoelectrons when (i) the intensity of incident light decreases, (ii) the target material is changed, (iii) the frequency of the incident light increases

The Einstein's photoelectric equation is

$$\left(\frac{1}{2}mv^2\right)_{max} = hv - W,$$

where v is the frequency of the incident light, W the work function of the metallic surface and $(½mv^2)_{max}$ the maximum kinetic energy (K_{max}) of the emitted electrons. If v_0 be the threshold frequency, then $W = hv_0$, and so

$$K_{max} = hv - hv_0.$$

A graph between K_{max} and v would be a straight line, cutting the v axis at v_0. The slope of this line is h.

(i) *Effect of Change in Light Intensity:* A or change in light intensity means a change in the number of photons falling per second on the metal. This will cause a corresponding change in the number of photoelectrons emitted. Thus if the light intensity is decreased, the number of photo electrons will decrease but their energy will remain the same. Hence the $K_{max} - v$ graph will remain unchanged.

(ii) *Effect of Change of Target Material:* The work function that must be done to take an electron through the emitting surface from just beneath it against the abaction of the rest of the metal. Obviously, a change in metal would change the work function W (and hence the threshold frequency v_0) which will change the energy of the photoelectrons. The new graph will be another straight line *parallel* to the previous line. Parallel, because slope (h) in fixed

(iii) *Effect of Change in Light Frequency:* A change in the frequency v of the incident light will cause a corresponding linear change in the maximum kinetic energy of the photoelectrons. A decrease in frequency (or increase in wavelength), for example, would decrease the energy. The number of photo electrons emitted per second will remain unaffected.

Importance of Compton Effect

When a *monochromatic* beam of X-rays is scattered by a substance, the scattered X-rays contain radiation not only of the same wavelength as that of the primary rays but also the radiation of longer wavelength. The former is called the 'unmodified radiation' and the latter the 'modified radiation'. The classical electromagnetic theory explained the unmodified radiation but it totally failed to explain the presence of the modified radiation. Compton, however, gave a satisfactory explanation for the modified radiation on the basis of quantum theory. Hence the phenomenon giving rise to modified radiation is called Compton effect.

Theory: According to the quantum theory, the primary X-ray beam is made up of photons of energy hv, where h is Planck's constant and v is the frequency of the primary X–rays. These photons travel with the speed of light a and possess momentum given by hv/c.

Compton assumed that during the scattering process, the incident photon collides with a free electron (assumed to be initially at rest in the laboratory coordinate system) in the scattering material (Fig.a). The photon transfers some of its energy to the electron which recoils with a velocity v in a direction making an angle ϕ (say) with the direction of the incident photon (Fig. b).

The photon itself *with reduced energy* is scattered in a direction θ with the original direction. These scattered photons constitute the scattered modified radiation. Let us apply the laws of conservation of energy and momentum to this collision.

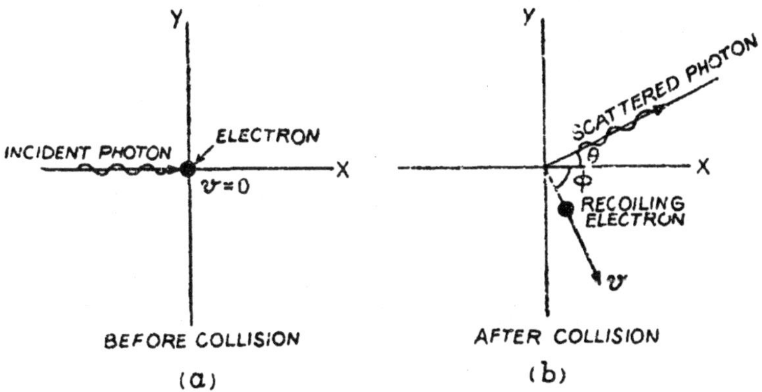

The energy and momentum of the incident photon are respectively $h\nu$ and $h\nu/c$.

The energy and momentum of scattered photon are respectively $h\nu'$ and $h\nu'/c$ where ν' is the frequency of the scattered radiation.

Since the recoiling electron may have a speed v which is comparable with the speed of light, we must use the relativistic expression for the energy and momentum of the electron. According to the theory of relativity, a mass m is equivalent to an amount of energy E by the relation

$$E = mc^2,$$

where c is the velocity of light. Further the mass of a particle varies with its velocity v according to the relation

$$m = \frac{m_0}{\sqrt{\left(1 - \frac{v^2}{c^2}\right)}}.$$

where m_0 is the mass of the particle when at rest. The total energy of the recoiling electron is the sum of its kinetic energy K plus its rest mass energy, *i.e.*

$$mc^2 = K + m_0 c^2.$$

Hence the kinetic energy of the electron is

$$K = mc^2 - m_0c^2$$

$$= \frac{m_0c^2}{\sqrt{\left(1-\frac{v^2}{c^2}\right)}} - m_0c^2$$

$$= m_0c^2\left[\frac{1}{\sqrt{\left(1-\frac{v^2}{c^2}\right)}} - 1\right]$$

Similarly, the momentum of the recoiled electron is

$$mv = \frac{m_0v}{\sqrt{\left(1-\frac{v^2}{c^2}\right)}}$$

Now, applying the law of conservation of energy, we get

$$hv = hv' + m_0c^2\left[\frac{1}{\sqrt{\left(1-\frac{v^2}{c^2}\right)}} - 1\right]. \quad \text{... (i)}$$

Again, applying the law of conservation of momentum along the X and Y axes, we have respectively

$$\frac{hv}{c} = \frac{hv'}{c}\cos\theta + \frac{m_0v}{\sqrt{\left(1-\frac{v^2}{c^2}\right)}}\cos\phi \quad \text{... (ii)}$$

and

$$0 = \frac{hv'}{c}\sin\theta + \frac{m_0v}{\sqrt{\left(1-\frac{v^2}{c^2}\right)}}\sin\phi \quad \text{... (iii)}$$

Re-arranging and squaring (*ii*) and (*iii*), we get

$$\left(\frac{hv}{c} - \frac{hv'}{c}\cos\theta\right)^2 = \frac{m_0^2\, v^2}{1-\dfrac{v^2}{c^2}}\cos^2\phi$$

and $$\left(\frac{hv'}{c}\sin\theta\right)^2 = \frac{m_0^2\, v^2}{1-\dfrac{v^2}{c^2}}\sin^2\phi$$

Adding: $$\left(\frac{hv}{c}\right)^2 + \left(\frac{hv'}{c}\right)^2 - \frac{2h^2vv'}{c^2}\cos\theta = \frac{m_0^2\, v^2}{1-\dfrac{v^2}{c^2}} \quad \text{... } (iv)$$

Re-arranging (*i*) and dividing by *c*, we get

$$\frac{hv}{c} - \frac{hv'}{c} + m_0c = \frac{m_0c}{\sqrt{\left(1-\dfrac{v^2}{c^2}\right)}}$$

Squaring:

$$\left(\frac{hv}{c}\right)^2 - \left(\frac{hv'}{c}\right)^2 + m_0^2c^2 - \frac{2h^2vv'}{c^2} + 2hm_0\left(v-v'\right) = \frac{m_0^2c^2}{1-\dfrac{v^2}{c^2}} \quad \text{... } (v)$$

Subtracting (*iv*) and (*v*), we get

$$m_0^2c^2 - \frac{2h^2vv'}{c^2}\left(1-\cos\theta\right) + 2hm_0\left(v-v'\right) = \frac{m_0^2}{1-\dfrac{v^2}{c^2}}\left(c^2-v^2\right) = m_0^2c^2$$

or $$2hm_0\,(v - v') = \frac{2h^2vv'}{c^2}\left(1-\cos\theta\right)$$

or $$m_0\,(v - v') = \frac{hvv'}{c^2}\left(1-\cos\theta\right)$$

Substituting $v = c/\lambda$ and $v' = c/\lambda'$ where λ and λ' are the wavelengths of the incident and scattered X-rays, we get

$$m_0c\left(\frac{1}{\lambda} - \frac{1}{\lambda'}\right) = \frac{h}{\lambda\lambda'}\left(1-\cos\theta\right)$$

or $$\lambda' - \lambda = \frac{h}{m_0 c}(1 - \cos\theta).$$

or $$\Delta\lambda = \frac{h}{m_0 c}(1 - \cos\theta).$$

This is the expression for the change in wavelength, $\Delta\lambda$ is called the 'Compton shift'.

Compton Wavelength: The quantity h/m_0c has dimensions of length, and is known as the 'Compton wavelength of the electron'. Its magnitude is

$$\frac{h}{m_0 c} = \frac{6.63 \times 10^{-34} \text{ joule-sec}}{\left(9.11 \times 10^{-31} \text{ kg}\right) \times \left(3.0 \times 10^{8} \text{ meter-sec}^{-1}\right)}$$

$$= 0.0243 \times 10^{-10} \text{ m} = 0.0243 \text{ Å}.$$

This shows that *the change in wavelength (Compton shift) depends neither on the incident wavelength nor on the scattering material,* but depends only on the angle of scattering.

At $$\theta = 90°;$$

$$\Delta\lambda = 0.0243 \text{ Å}.$$

This result was verified by Compton himself.

Presence of the Unmodified Radiation: The above theory holds for the 'free' or 'loosely-bound' electrons of the atom. If the photon strikes an electron '*tightly bound*' to the atom, then the hole atom, and not simply the electron, recoils. Since the mass of the atom much larger, the Compton shift $\frac{h}{m_0 c}$ $(1 - \cos\theta)$, m_0 now being the mass of the atom, is negligible. This accounts for the unmodified radiation. In the heavier elements in which most of the electrons are tightly bound, the modified radiation is too weak.

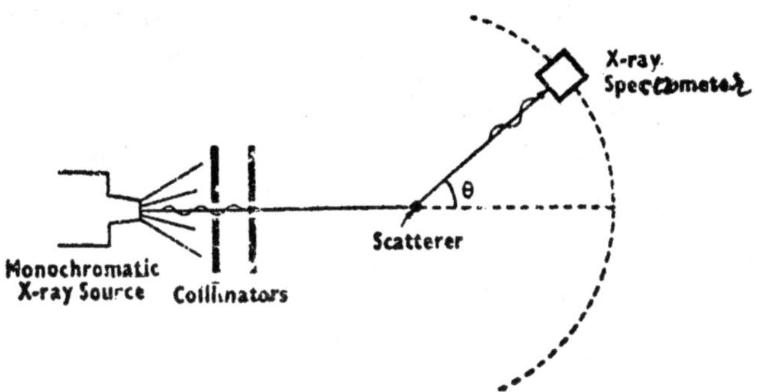

The Compton effect offers one of the strongest evidences in support of the Planck's quantum theory of radiation.

Experimental Verification: A beam of monochromatic X-rays of known wavelength λ is made to fall on a graphite scattered. The distribution of intensity with wavelength in the X-rays scattered at various angles θ is measured by means of a Bragg's X-ray spectrometer (Fig.). The results have been shown in. Fig. For $\theta = 90°$, $\Delta\lambda$ is 0.0242 Å, which is an agreement with Compton formula.

Measurement of Energy of the Recoiled Electrons: From eq. (*ii*), in which $v' < v$, it is clear that necessarily $\cos \phi > 0$ and hence $\phi < 90°$, so that the electron must move in the forward direction. These recoiled electrons, whose existence was predicted by Compton, were experimentally discovered by C. T. R. Wilson and Bothe. Bless measured the energies for these electrons recoiled in various directions with various velocities by using a magnetic spectrogram. The electrons are subjected to a magnetic; field in which they describe circles and are received on a photographic plate. Those electrons which leave different parts of the scanner with the same velocity describe circles of same radius and are focussed on the same line on the plate. By measuring the positions of the various lines on the plate (which represent groups of recoiled electrons), the radii of the corresponding circles are known from which the velocities are computed. Bless found that the results were in agreement with Compton's predictions.

A photon of energy hv is scattered through an angle θ by a free electron originally at rest. Below it is shown that the ratio of the kinetic energy of the recoil electron to the energy of the incident photon is $\frac{x(1-\cos\theta)}{1+x(1-\cos\theta)}$ ***where*** $x = \frac{hv}{m_0c^2}$.

The maximum possible kinetic energy of the recoiling electron in terms of the incident photon energy may be computed.

The energy of incident photon is

$$E = hv.$$

Let v' be the frequency of the scattered photon and K the kinetic energy of the recoil electron. By conservation of energy, we have

$$hv = hv' + K.$$

$$\therefore \qquad \frac{K}{E} = \frac{hv - hv'}{hv} = \frac{(hv/\lambda) - (hv/\lambda')}{hc/\lambda}$$

$$= \frac{\lambda' - \lambda}{\lambda'} = \frac{\Delta\lambda}{\lambda + \Delta\lambda}.$$

The Compton shift $\Delta\lambda = \frac{m}{m_0c}(1-\cos\theta)$.

$$\therefore \qquad \frac{K}{E} = \frac{(h/m_0c)(1-\cos\theta)}{\frac{c}{v} + \frac{h}{m_0c}(1-\cos\theta)}$$

$$= \frac{\frac{hv}{m_0c^2}(1-\cos\theta)}{1 + \frac{h}{m_0c^2}(1-\cos\theta)}$$

$$= \frac{x(1-\cos\theta)}{1+x(1-\cos\theta)} \quad \text{where } x = h\upsilon/m_0c^2.$$

The last equation may be written as

$$\frac{K}{E} = \frac{2x\sin^2\frac{\theta}{2}}{1+2x\sin^2\frac{\theta}{2}}$$

$$K = \frac{E}{\dfrac{1}{2x\sin^2\frac{\theta}{2}}+1}$$

K is maximum when the denominator on the R.H.S. is minimum or $\sin^2\frac{\theta}{2}$ is maximum, *i.e.* $\theta = 180°$.

$$\therefore \qquad K_{max} = \frac{E}{\dfrac{1}{2x}+1} = \frac{2\,Ex}{1+2x}.$$

Now $\qquad E = h\upsilon$ and $x = h\upsilon/m_0c^2$.

$$\therefore \qquad K_{max} = \frac{2h^2\upsilon^2/m_0c^2}{1+2h\upsilon/m_0c^2}.$$

Comparison of Compton Effect and Photoelectric Effect: Let us consider a Compton collision between a photon and a free electron (initially at rest). The photon has an energy $h\upsilon$ and a momentum $h\upsilon/c$. Suppose the photon transfers its entire energy (and momentum) to the electron who acquires a velocity υ. Then, from consideration of the conservation of energy and momentum, we have

$$h\upsilon = \tfrac{1}{2}m\upsilon^2$$

and $$\frac{h\upsilon}{c} = m\upsilon,$$

where m is the mass of the electron. Eliminating $h\nu$ between these two equations we find that the velocity of the electron comes out to be

$$v = 2c,$$

that is, twice the velocity of light. This is impossible according to the relativistic mechanics. Hence we conclude that a *'free' electron cannot absorb 'all' the energy of a photon and conserve both energy and momentum.*

In photoelectric effect, an electron absorbs the incident photon *completely* before it leaves the photo-sensitive surface as a photoelectron. Obviously, this electron *cannot be free, i.e.* it must be bound to an atom otherwise it could not absorb the photon completely. Hence we conclude that the photoelectric effect cannot occur for completely free electrons; the electrons must be bound to an atom.

Thus there is an important difference between the photoelectric effect and the Compton effect. In the photoelectric effect the photon completely disappears and all of its energy is given to the photoelectron. In Compton effect there is still a photon after the collision but its energy is less than that of the incident photon, only a part of the energy of the incident photon has been given to the Compton electron. Thus in Compton effect, the electron can be free.

Compton Effect not Observed for Visible Light: The change in wavelength due to the Compton effect is given by

$$\Delta\lambda = \frac{h}{m_0 c}(1 - \cos\theta) = 0.0243\,(1 - \cos\theta)\ \text{Å},$$

where θ is the scattering angle. Obviously, the change $\Delta\lambda$ is independent of the wavelength of the radiation suffering Compton effect, and depends only on the scattering angle θ. The greatest value of $(1 - \cos\theta)$ is 2 when $\theta = 180°$, so that the maximum wavelength-change possible is 0.0486 or roughly

0.05 Å only. This means that Compton effect can be detected only for those radiations whose wavelength is not greater than a few λ. For example, for λ = 5 Å, there is a maximum wavelength-change of 1 %, while for λ = 1 Å there is a 5% change. For visible light (λ ≃ 5000 Å) the maximum wavelength change (0.05 Å) is only about 0.001% of the initial wavelength which is undetectable.

PROBLEMS

1. The photoelectric work function of a metal is 2.061 eV. Calculate the threshold wavelength and frequency for the metal. (h = 6.6 × 10^{-34} joule-sec, e = 3.0 × 10^{8} m/sec, 1 eV = 1.6 × 10^{-19} joule).

Solution: A radiation photon cannot emit photo electron from a metal if its energy is less than the work function W of the metal. If v_0 be the threshold frequency of radiation, the threshold energy of the photon will be hv_0. Thus

$$W = hv_0.$$

$$\therefore \qquad v_0 = \frac{W}{h} = \frac{2.061 \times 1.6 \times 10^{-19}\,\text{joule}}{6.6 \times 10^{-34}\,\text{joule-sec}}$$

$$= 5.0 \times 10^{14}\ \text{sec}^{-1}.$$

The threshold wavelength is

$$\lambda_0 = \frac{c}{v_0} = \frac{3.0 \times 10^{8}\ \text{m-sec}^{-1}}{5.0 \times 10^{14}\,\text{sec}^{-1}}$$

$$= 6 \times 10^{-7} = 6000\ \text{Å}.$$

2. The work function of sodium is 2.3 eV. Let's see what is the maximum wavelength of light that will cause the photoelectrons to be ejected. It would be eager to look into whether sodium shows a photoelectric effect for orange light, λ = 6800 Å (h = 6.626 × 10^{-34} J-s, c = 3.0 × 10^{8} m/s).

Its answer is: 5400 Å, No)

3. The work function in electron-volts for a metal, given that the photoeleciric threshold wavelength is 6800 Å is calculated below

Solution: Threshold wavelength λ_0 = 6800 Å = 6800 × 10^{-10} meter. If v_0 be the threshold frequency, then the work function for the metal is

$$W = hv_0 = \frac{hc}{\lambda_0}$$

$$= \frac{(6.6 \times 10^{-34} \text{ joule-sec}) \times (3.0 \times 10^8 \text{ m-sec}^{-1})}{2800 \times 10^{-10} \text{m}}$$

$$= 2.9 \times 10^{-19} \text{ joule.}$$

Now $1 \text{ eV} = 1.6 \times 10^{-19}$ joule.

$$\therefore \quad W = \frac{2.9 \times 10^{-19}}{1.6 \times 10^{-19}} = 1.8\text{eV}.$$

4. A photon of wavelength 3310 Å falls on a photo cathode and ejects an electron of maximum energy 3 × 10^{-12} erg. The work function of the cathode material is calculated here. (h = 6.62 × 10^{-27} erg-sec, c = 3 ×10^{10} cm/sec and 1 eV = 1.6 × 10^{-12} erg).

Solution: When a photon of frequency v falls on a photocathode of work function W, the maximum kinetic energy K_{max} of the photoelectrons is given by the Einstein photoelectric equation:

$$K_{max} = hv - W$$

or

$$W = hv - K_{max} = \frac{hc}{\lambda} - K_{\text{max}}$$

Substituting the given values, we get

$$W = \frac{(6.62 \times 10^{-27} \text{ erg-sec}) \times (3 \times 10^{10} \text{cm-sec}^{-1})}{(3310 \times 10^{-8} \text{ cm})} - (3 \times 10^{-12} \text{erg})$$

$$= (6 \times 10^{-12}\ \text{erg}) - (3 \times 10^{-12}\ \text{erg}) = 3 \times 10^{-12}\ \text{erg}$$

$$= \frac{3 \times 10^{-12}}{1.6 \times 10^{-12}} = 1.875\ \text{eV}.$$

5. The wavelength of the photoelectric threshold of a metal is 2300 Å. Let's determine (i) the work function in eV and (ii) the maximum kinetic energy (in eV) of the photoelectrons ejected by ultraviolet light of wavelength 1800 Å. (h = 6.626 × 10^{-34} joule-sec).

Solution: (i) Threshold wavelength λ_0 = 2300Å = 2300 × 10^{-10} meter. If v_0 be the threshold frequency, then the work function of the surface is

$$W = hv_0 = \frac{hv}{\lambda_0}$$

$$= \frac{(6.626 \times 10^{-34}\ \text{joule-sec}) \times (3 \times 10^{8}\ \text{m-sec}^{-1})}{2300 \times 10^{-10}\ \text{m}}$$

$$= 8.64 \times 10^{-19}\ \text{joule}$$

$$= \frac{8.64 \times 10^{-19}}{1.6 \times 10^{-19}} = 5.4\ \text{eV}.[\because 1\ \text{eV} = 1.6 \times 10^{-19}\ \text{joule}]$$

(*ii*) Let K_{max} be the kinetic energy of the fastest photoelectrons when light of frequency v falls on the surface. Then, from the Einstein's photoelectric equation, we have

$$K_{max} = hv - W.$$

If λ be the incident wavelength, we have

$$K_{max} = \frac{hc}{\lambda} - W.$$

$$\text{Here:} \frac{hc}{\lambda} = \frac{(6.626 \times 10^{-34}\ \text{joule-sec}) \times (3 \times 10^{8}\ \text{m-sec}^{-1})}{(1800 \times 10^{-10}\ \text{m})}$$

$$= 11.0 \times 10^{-19}\ \text{joule} = \frac{11.0 \times 10^{-19}}{1.6 \times 10^{-19}} = 6.9\ \text{eV}.$$

and W = 5.4 eV (as obtained above).

$$\therefore \quad K_{max} = \frac{hc}{\lambda} - W = 6.9 - 5.4 - 1.5 \text{ eV}.$$

6. The threshold wavelength for photoelectric emission from a metallic surface is 5800Å. Let's find the maximum energy of the photoelectrons if the wavelength of the incident light is 4500Å. (h = 66 × 10^{-34} joule-sec, c = 3 × 10^{8} m/sec).

The answer is 0.62 eV.

7. The threshold frequency for photoelectric emission in copper is 1.1 × 10^{15} sec^{-1}. The maximum energy of the photoelectrons when light of frequency 15 × 10^{16} sec^{-1} falls on copper surface may be foundout. Let's also calculate the retarding potential. (h = 6.62 × 10^{-34} joule-sec)

Solution: If v_0 be the threshold frequency, then according to Einstein's equation, the maximum kinetic energy of the photoelectrons is

$$K_{max} = hv - hv_0$$

$$= 6.62 \times 10^{-34} (1.5 \times 10^{15} - 1.1 \times 10^{15})$$

$$= 2.648 \times 10^{-19} \text{ joule}.$$

But 1 eV = 1.6 × 10^{-19} joule.

$$\therefore \quad K_{max} = \frac{2.648 \times 10^{-19}}{1.6 \times 10^{-19}} = 1.655 \text{ eV}.$$

The retarding potential V_0 is the potential which just stops the photoelectrons of maximum kinetic energy. Thus

$$V_0 = 1.655 \text{ volts}.$$

8. The work function of a metal surface is 12 eV. The kinetic energy of the fastest and slowest photoelectrons and the retarding potential when light of frequency 5.5 × 10^{14} sec^{-1} falls on the surface is calculated below. (h = 662 × 10^{-34} joule-sec).

Solution: Let K_{max} be the kinetic energy of the fastest photoelectrons. Then the photoelectric equation is

$$K_{max} = hv - W, \qquad \text{... (i)}$$

where W is the work function and v the incident frequency. Let us evaluate hv in electron-volts. This is

$$hv = (6.62 \times 10^{-34} \text{ joule-sec})(5.5 \times 10^{14} \text{ sec}^{-1})$$

$$= 3.64 \times 10^{-19} \text{ joule.}$$

But $\quad 1 \text{ eV} = 1.60 \times 10^{-19}$ joule.

$$\therefore \quad hv = \frac{6.34 \times 10^{-19}}{1.60 \times 10^{-19}} = 2.27 \text{ eV.}$$

Substituting this value of hv and the given value of W (= 1.2 eV) in eq (*i*), we get

$$K_{max} = 2.27 - 1.2 = 1.07 \text{ eV.}$$

The kinetic energy of the slowest photoelectrons is zero.

The stopping potential V_0 is measured by the kinetic energy K_{max} of the fastest photoelectrons which is 1.07 electron-volts.

$$\therefore \quad V_0 = 1.07 \text{ volts.}$$

9. The stopping potential for photoelectrons emitted from a surface illuminated by light of wavelength 5893 Å is 0.36 volt. Let's calculate the maximum kinetic energy of photoelectrons, the work function of the surface and the threshold frequency. (h = 6.62 × 10^{-34} joule-sec, c = 3.0 × 10^8 m/sec and 1 eV = 1.6 × 10^{-19} joule.)

Solution: If V_0 volt be the stopping potential, then the (maximum) kinetic energy of the fastest electrons is given by

$$K_{max} = V_0 \text{ electron-volt} = 0.36 \text{ eV.}$$

Let W be the work function of the surface illuminated by

light of frequency v (or wavelength λ). Then from the Einstein's photoelectric equation, we have

$$K_{max} = hv - W = \frac{hc}{\lambda} - W,$$

$$\therefore \quad W = \frac{hc}{\lambda} - K_{max}$$

Substituting the given values, we have

$$W = \frac{(6.62 \times 10^{-34} \text{ joule-sec}) \times (3 \times 10^{8} \text{ m/sec})}{5893 \times 10^{-10} \text{ m}}$$

$$-(0.36 \times 1.6 \times 10^{-19} \text{ Joule})$$

$$= 3.37 \times 10^{-19} \text{ joule} - 0.576 \times 10^{-19} \text{ joule}$$

$$= 2.794 \times 10^{-19} \text{ joule}$$

$$= \frac{2.794 \times 10^{-19}}{1.6 \times 10^{-19}} = 1.746 \text{ eV}.$$

The threshold frequency is

$$v_0 = \frac{W}{h} = \frac{2.794 \times 10^{-19} \text{ joule}}{6.62 \times 10^{-34} \text{ joule-sec}}$$

$$= 4.22 \times 10^{14} \text{ hertz.}$$

10. A radiation of frequency 10^{16} Hz falls on a photocathode and ejects electrons with maximum energy of 4.2×10^{-19} J. If the frequency of radiation is changed to 5×10^{14} Hz, the maximum energy of ejected electrons becomes 09×10^{-19} J. Planck's constant, threshold frequency and work function of photo-cathode material is calculated below.

Solution: The photoelectric equation is

$$K_{max} = hv - W,$$

where v is the incident frequency and W is the work function.

Substituting the given values in the two cases, we have

$$4.2 \times 10^{-19} \text{ J} = h\,(10 \times 10^{14} \text{ s}^{-1}) - W \qquad \text{... } (i)$$

and $$0.9 \times 10^{-19} \text{ J} = h\,(5 \times 10^{14} \text{ s}^{-1}) - W. \qquad \text{... } (ii)$$

Subtracting eq. (*ii*) from eq. (*i*), we get

$$3.3 \times 10^{-19} \text{ J} = h\,(5 \times 10^{14} \text{ s}^{-1})$$

or $$h = \frac{3.3 \times 10^{-19} \text{ J}}{5 \times 10^{14} \text{ s}^{-1}} = 6.6 \times 10^{-34} \text{ J-s}$$

Putting this value of h in eq. (*i*) and solving, we get

$$W = 2.4 \times 10^{-19} \text{ J}.$$

The threshold frequency is

$$v_0 = \frac{W}{h} = \frac{2.4 \times 10^{-19} \text{ J}}{6.6 \times 10^{-34} \text{ J-s}} = 3.6 \times 10^{14} \text{ s}^{-1}.$$

11. When tungsten surface is illuminated with light of wavelength 1800 Å, the maximum kinetic energy of the electrons liberated is 1.5 eV. Photoelectric emission ceases when the wavelength of the incident light exceeds 2300Å. As for this case Planck's constant is calculated below ($c = 3.0 \times 10^8$ m/s, 1 eV = 1.6×10^{-19} J).

The maximum kinetic energy of photoelectrons emitted from a surface of work function W is given by

$$K_{max} = hv - W, \qquad \text{... } (i)$$

where v is the frequency of incident light.

For $$\lambda = 1800 \text{ Å},\ v = \frac{c}{\lambda} = \frac{3.0 \times 10^8 \text{ m/s}}{1800 \times 10^{-10} \text{ m}}$$

$$= 1.66 \times 10^{15} \text{ sec}^{-1}$$

and $$K_{max} = 1.5 \text{ eV} = 1.5 \times 1.6 \times 10^{-19}$$

$$= 2.4 \times 10^{-19} \text{ J}.$$

For $\lambda = 2300\text{Å}$,

$$v = \frac{3.0 \times 10^8 \text{ m/s}}{2300 \times 10^{-10} \text{ m}}$$

$$= 1.30 \times 10^{15} \text{ sec}^{-1}.$$

and $K_{max} = 0$

Substituting these values in eq. (*i*), we get

$$2.4 \times 10^{-19} \text{ J} = h(1.66 \times 10^{15} \text{ s}^{-1}) - W. \quad \text{... (ii)}$$

and
$$0 = h\,(1.30 \times 10^{15} \text{ s}^{-1}) - W. \quad \text{... (iii)}$$

Subtracting eq. (*iii*) from eq (*ii*), we get

$$2.4 \times 10^{-19} \text{ J} = \text{h}\,(0.36 \times 10^{16} \text{ s}^{-1})$$

or
$$h = \frac{2.4 \times 10^{-19} \text{ J}}{0.36 \times 10^{15} \text{ s}^{-1}}$$

$$= 6.66 \times 10^{-34} \text{ J-s}.$$

12. *Light of wavelength 3000 Å falls on a metal surface having a work function of 2.3 eV. The maximum velocity of the ejected electrons is calculated as follows. (h = 6.6 × 10⁻³⁴ joule-sec, c = 3.0 × 10⁸ m/sec, electronic mass m = 9.1 × 10⁻³¹ kg and 1 eV = 1.6 × 10⁻¹⁹ joule).*

Solution: The photoelectric equation is

$$K_{max} = hv - W = \frac{hc}{\lambda} - W$$

$$= \frac{(6.6 \times 10^{-34} \text{ joule sec})\,(3.0 \times 10^{8} \text{ m sec}^{-1})}{3000 \times 10^{-10} \text{ m}}$$

$$- (2.3 \times 1.6 \times 10^{-19} \text{ joule})$$

$$= 6.6 \times 10^{-19} \text{ joule} - 3.7 \times 10^{-19} \text{ joule}$$

$$= 2.9 \times 10^{-19} \text{ joule}.$$

Now, $K_{max} = ½\, mv_{max}^{2}$

$\therefore$ $v_{max} = \sqrt{\left(\frac{2K_{max}}{m}\right)}$

$= \sqrt{\left(\frac{2 \times 2.9 \times 10^{-19}}{9.1 \times 10^{-31}}\right)} = 8 \times 10^{5}$ m/sec.

13. The work function of a metal is 2.30 eV. The maximum wavelength of light which can liberate photo electrons from it, and the maximum velocity of the photoelectrons emitted by light of wavelength 2537 Å may be calculated as follows. ($h = 6.62 \times 10^{-27}$ erg-sec, $c = 3.0 \times 10^{10}$ cm/sec, electronic mass $m = 9.1 \times 10^{-28}$ gm and 1 eV = 1.6×10^{-12} erg).

The answer is 5400 Å, 9.54×10^{7} cm/sec.

14. A beam of X-rays is scattered by free electrons. At 45° from the beam direction the scattered X-rays have a wavelength of 0.022 Å. The wavelength of the X-rays in the direct beam is given below.

Solution: The Compton shift (increase in wavelength) is given by

$$\Delta\lambda = \frac{h}{m_0 c}(1 - \cos\theta)$$

$$= 0.0243\ (1 - \cos\theta)\ \text{Å}.$$

For $\theta = 45°$, we have

$$\Delta\lambda = 0.0243\ (1 - 0.707) [\because \cos 45° = 0.707]$$

$$= 0.0071\ \text{Å}.$$

The scattered wavelength is $\lambda = 0.022$ Å. Therefore, the incident wavelength must be

$$\lambda = \lambda' - \Delta\lambda$$

$$= 0.022 - 0.0071 = 0.0149\ \text{Å}.$$

15. The maximum change in wavelength in a Compton scattering is calculated as follows.

Solution: The Compton change in wavelength is given by

$$\Delta \lambda = \frac{h}{m_0 c}(1 - \cos\theta)$$

$$= 0.0243\ (1 - \cos\theta)\ \text{Å}.$$

The change in wavelength is maximum when the photon recoils back, *i.e.* when $\theta - 180°$. Then we have

$$(\Delta\lambda)_{max} = 0.0243\ (1 - \cos 180°) = 0.0486\ \text{Å}.$$

16. X-rays of wavelength 1.0 Å are scattered at such an angle that the recoil electron has the maximum kinetic energy. The wavelength of the scattered ray and the energy of recoil electrons is calculated as follows. h = 6.63 × 10^{-34} J–s, c = 3 × 10^8 m/s.

Solution: The recoil electron acquires maximum kinetic energy when the Compton shift is maximum (or the X-rays are scattered at 180°).

The Compton shift (change in wavelength) is given by

$$\Delta \lambda = 0.0243\ (1 - \cos\theta)\ \text{Å}.$$

It is maximum when θ = 180°, *i.e.* $\cos\theta = -1$.

Thus

$$(\Delta \lambda)_{max} = 0.0486\text{Å}.$$

The incident wavelength is λ = 1.0Å. Therefore, the scattered wavelength will be

$$\lambda' = \lambda + \Delta\lambda = 1.0 + 0.0486 = 1.0486\ \text{Å}.$$

The energy of the incident X-ray photon is $h\nu$ (= hc/λ) and that of the scattered photon is $h\nu'$ (= hc/λ'). The balance has been imparted to the recoiling electron. Let it be K.

Then

$$K = \frac{hc}{\lambda} - \frac{hc}{\lambda'}$$

$$= \frac{hc(\lambda' - \lambda)}{\lambda\lambda'} = \frac{hc\,\Delta\lambda}{\lambda\lambda'}.$$

Substituting the values of h, c, λ, λ' and $(\Delta\lambda)_{max}$, the maximum kinetic energy is

$$K = (6.63 \times 10^{-34} \text{ joule sec}) \times (3 \times 10^{8} \text{ meter-sec}^{-1}) \times \frac{(0.0486 \times 10^{-10} \text{ meter})}{(1.0 \times 10^{-10} \text{ meter}) \times (1.0486 \times 10^{-10} \text{ meter})}$$

$$= 0.92 \times 10^{-16} \text{ joule}$$

$$= \frac{0.92 \times 10^{-16}}{1.6 \times 10^{-19}} = 575 \text{ eV}.$$

17. X-rays of λ = 1.0 Å are scattered at 60° from the free electrons. After calculating the shift in wavelength and the energy of the recoil electrons. The answer is 0.012Å, 147 eV.

An X-ray photon is found to have its wavelength doubled on being scattered through 90°. The wavelength and energy of the incident photon will be as follows. (m_0 of electron = 9.0 × 10^{-28} gm).

Solution: The Compton change in wavelength is given by

$$\Delta\lambda = \frac{h}{m_0 c}(1 - \cos\theta).$$

When θ = 90°, $\Delta\lambda = \lambda$ (wavelength is doubled). Thus

$$\lambda = \frac{h}{m_0 c} = 0.0243 \text{ Å}$$

This is the wavelength of the incident photon. Its energy would be

$$E = h\upsilon = \frac{hc}{\lambda}$$

$$= \frac{hc}{h/m_0c} = m_0c^2$$

$$= (9.0 \times 10^{-28} \text{ gm}) \times (3.0 \times 10^{-28} \text{ cm/sec})^2$$

$$= 8.1 \times 10^{-7} \text{ erg.}$$

18. It is interested to find answer as to for what wavelength photon does Compton scattering result in a photon whose energy is one-half that of the original, at a scattering angle of 45° and in which region of electromagnetic spectrum does such a photon lie (Compton wavelength of the electron = 0.0242Å). Solution is found as follows

The Compton change in wavelength at 45° scattering is given by

$$\Delta\lambda = \frac{h}{m_0c}(1 - \cos 45°)$$

$$= 0.0242\text{Å } (1 - 0.707)$$

$$= 0.0071 \text{ Å}.$$

The energy of the incident photon is hc/λ and that of the scattered photon is $hc/(\lambda + \Delta\lambda)$. Here it is given that

$$\frac{hc}{\lambda + \Delta\lambda} = \frac{1}{2}\frac{hc}{\lambda}$$

or $$\lambda = \Delta\lambda = 0.0071 \text{ Å}$$

This wavelength lies in the γ-region of the electromagnetic spectrum.

19. X-rays of wavelength 1.0 Å are scattered from a carbon block. Let's find (i) the wavelength of the scattered beam in a direction making 90° with the incident beam, (ii) kinetic

energy imparted to the recoiling electron, (iii) direction of the recoiling electron, (h = 6.63 × 10⁻³⁴ joule-sec, c = 3.0 × 10⁸ meters/sec and 1 eV = 1.6 × 10⁻¹⁹ joule).

Solution: (*i*) The Compton shift is

$$\Delta \lambda = 0.0243(1\text{-cose } \theta) \text{ Å}.$$

At $\theta = 90°$, we have

$$\Delta \lambda = 0.0243\ (1 - \cos 90°) = 0.0243\text{Å}$$

Hence the wavelength of the scattered beam is

$$\lambda' = \lambda + \Delta \lambda = 1.0 + 0.0243 = 1.0243\text{Å}$$

(*ii*) The energy of the incident *X*-ray photon is $h\nu$ ($= hc/\lambda$) and that of the scattered photon is $h\nu'$ ($= hc/\lambda'$). The balance has been imparted to the recoiling electron. Let it be *K*. Then

$$K = \frac{hc}{\lambda} - \frac{hc}{\lambda'}$$

$$= \frac{hc\left(\lambda' - \lambda\right)}{\lambda\lambda'} = \frac{hc\,\Delta\lambda}{\lambda\lambda'}$$

Substituting the values of h, c, λ, λ' and $\Delta\lambda$, we get

$$K = (6.63 \times 10^{-34} \text{ joule-sec})$$

$$\times\ (3 \times 10^{8} \text{ meter-sec}^{-1})$$

$$\times \frac{\left(0.0243 \times 10^{-10} \text{ meter}\right)}{\left(1.0 \times 10^{-10} \text{ meter}\right) \times \left(1.0243 \times 10^{-10} \text{meter}\right)}$$

$$= 4.72 \times 10^{-17} \text{ joule}$$

$$= \frac{4.72 \times 10^{-17}}{1.6 \times 10^{-19}} = 295 \text{ eV}.$$

(*iii*) Suppose the recoiling electron appears in a direction making an angle ϕ with the direction of the incident photon.

Applying conservation of momentum along and perpendicular to the direction of the incident photon, we get

$$\frac{hv}{c} = \frac{hv'}{c} \cos 90^\circ + mv \cos \phi$$

and
$$0 = \frac{hv'.}{c} \sin 90^\circ - mv \sin \phi.$$

Dividing, we get

$$\tan \phi = \frac{v'}{v} = \frac{\lambda}{\lambda'}$$

Substituting the values of λ and λ, we get

$$\tan \phi = \frac{1.0}{1.0243} = 0.98$$

$$\therefore \quad \phi = \tan{-1}\ (0.98) = 44^\circ.$$

20. ***Let's in Compton effect, if the incident photon has wavelength 2.0 × 10^{-8} cm and $\theta = 90^\circ$ deduce (i) the wavelength of the scattered photon, (ii) the energy of the recoil electron, (ii) the angle at which the recoil electron appears. (A= 6.62 ×10^{-27} erg-sec, c = 3.0 × 10^{10} cm-sec^{-1}, 1 eV = l.6 × 10^{-12} erg).***

Answer is (i) 2.0243Å, (ii) 74.5 eV, (iii) $\tan^{-1}$ (0.99).

21. ***In Compton scattering the incident radiation has a wavelength of 2A, and that scattered through 180° has 2.048 Å. The energy of recoil electron which scatters radiation through 60°. (Values of A, c and 1 eV same as above) may be calculated.***

The answer is 37 eV.

22. ***Photons of wavelength 0.0124Å are scattered by free electrons at angles 90° and 180′. The wavelengths of the scattered photons and the energies transferred to the free electrons in the two cases found out as follows (A = 6.62 × 10^{-34} joule-sec, c - 3.0 × 10^8 m/sec, m_0 = 9.1 × 10^{-31} kg and 1 eV = 1.6 × 10^{-19} joule).***

Solution: The Compton-increase in wavelength is

$$\Delta\lambda = \frac{h}{m_0 c} (1 - \cos\theta) = 0.0243 (1 - \cos\theta) \text{ Å}$$

For $\theta = 90°$;

$$\Delta\lambda = 0.0243 (1 - \cos 90°) = 0.0243 \text{ Å}.$$

Therefore, the scattered wavelength is

$$\lambda' = \lambda + \Delta\lambda = 0.0124 + 0.0243 = 0.0367 \text{ Å}.$$

The kinetic energy transferred to the free electron is

$$K = \frac{hc\,\Delta\lambda}{\lambda\lambda'}$$

$$= \frac{(6.62 \times 10^{-34} \text{ joule-sec})(3.0 \times 10^{8} \text{ m/sec})(0.0243 \times 10^{-10} \text{ m})}{(0.0124 \times 10^{10} \text{ m})(0.0367 \times 10^{-10} \text{ m})}$$

$$= 1.06 \times 10^{-13} \text{ joule}$$

$$= \frac{1.06 \times 10^{-13}}{1.6 \times 10^{-19}} = 0.66 \times 10^{6} \text{ eV} = 0.66 \text{ MeV}.$$

For $\theta = 180°$; we have

$$\Delta\lambda = 0.0243 (1 - \cos 180°) = 0.0486 \text{ Å}.$$

Therefore, the scattered wavelength is

$$\lambda' = \lambda + \Delta\lambda = 0.0124 + 0.0486 = 0.0610 \text{ Å}.$$

The kinetic energy transferred to the free electron is

$$K = \frac{hc\,\Delta\lambda}{\lambda\lambda'}$$

$$= 1.28 \times 10^{-13} \text{ joule} \qquad \text{(as calculated above)}$$

$$= 0.80 \text{ MeV}.$$

24. A beam of γ-radiation having photon energy of 510 keV is incident on a foil of aluminium. The wavelength of the radiation scattered at 90° and also the energy and direction of emission of the recoil electron may be calculated.

Solution: The energy of the incident y-radiation is

$$E = 510 \text{ ke V} = 510 \times 10^3 \text{ eV}$$

$$= 510 \times 10^3 \times 1.6 \times 10^{-19} = 8.16 \times 10^{-14} \text{ joule.}$$

Its wavelength is given by

$$\lambda = \frac{hc}{E}$$

$$= \frac{(6.62 \times 10^{-34} \text{ joule-sec}) \times (3.0 \times 10^8 \text{ m/sec})}{8.16 \times 10^{-14} \text{ joule}}$$

$$= 2.43 \times 10^{-12} \text{ meter} = 0.0243\text{Å}.$$

The Compton-shift in wavelength at scattering angle of 90° is given by

$$\Delta \lambda = 0.0243 \text{ (l-cos } 90°) = 0.0243\text{Å}.$$

Hence the wavelength of the scattered radiation is

$$\lambda' = \lambda + \Delta \lambda = 0.0243 + 0.0243 = 0.0486 \text{ Å}.$$

The kinetic energy of the recoiling electron is given by

$$K = \frac{hc\,\Delta\lambda}{\lambda\lambda'}$$

$$= \frac{(6.62 \times 10^{-34} \text{ joule-sec})(3.0 \times 10^8 \text{ m/sec})(0.0243 \times 10^{-1} \text{ m})}{(0.0243 \times 10^{-10}\text{m})(0.0486 \times 10^{-10}\text{m})}$$

$$= 4.1 \times 10^{-14} \text{ joule} = \frac{4.1 \times 10^{-14}}{1.6 \times 10^{-19}} = 2.56 \times 10^5 \text{ eV}$$

$$= 256 \text{ KeV}$$

The direction ϕ of the recoil electron *for 90° scattering* is given by

$$\tan \phi = \frac{v'}{v} = \frac{\lambda}{\lambda'} = \frac{0.0243}{0.0486} = 0.5$$

$$\therefore \ \phi = \tan^{-1} (0.5) = 26.5°.$$

25. *y-rays of energy 0.88 MeV are made to fall on a sheet of aluminium. Calculate the maximum energy of Compton recoil electrons is calculated as follows. (h = 6.62 ×10^{-34} joule-sec, c = 30 ×10^{8} m/sec and 1 eV = 1.6 × 10^{-18} joule).*

Solution: The wavelength of the y-rays of energy 0.88 MeV is given by

$$\lambda = \frac{hc}{E}$$

$$= \frac{(6.62 \times 10^{-34} \text{ joule-sec})(3.0 \times 10^{8} \text{ m/sec})}{(0.88 \times 10^{6} \times 1.6 \times 10^{-19} \text{ joule})}$$

$$= 14.1 \times 10^{-13} \text{ m} = 0.0141 \text{ Å}.$$

The Compton wavelength shift is given by

$$\Delta \lambda = 0.0243 \,(1—\cos \theta) \text{ Å}.$$

It is maximum for θ — 180°, and is given by

$$(\Delta \lambda)_{max} = 0.0486 \text{ Å}.$$

The corresponding scattered wavelength is

$$\lambda' = 0.0141 + 0.0486 = 0\ 0627 \text{ Å}.$$

The energy of recoil electrons is maximum for maximum Compton shift and is given by

$$K_{max} = \frac{hc(\Delta\lambda)_{max}}{\lambda\lambda'}$$

$$= \frac{(6.62 \times 10^{-34} \text{ joule-sec})(3.0 \times 10^{8} \text{ m/sec})(0.0486 \times 10^{-10} \text{ m})}{(0.0141 \times 10^{-10} \text{ m})(0.0627 \times 10^{-10} \text{ m})}$$

$$= 1.092 \times 10^{-13} \text{ joule}$$

$$= \frac{1.092 \times 10^{-13}}{1.6 \times 10^{-19}}$$

$$= 0.682 \times 10^{6} \text{ eV} = 0.682 \text{ MeV}$$

26. A monochromatic X-ray beam strikes a metal and a registers the photons. The adjacent detector scattered at 90°. The adjacent figure shows the energy spectrum of the photons observed at 90°. The parts which area modified due to Compton scattering is explained below.

Solution: The energy of the scattered photon is less than the energy of the incident photon. Hence the first peak, which corresponds to reduced energy is modified due to Compton scattering.

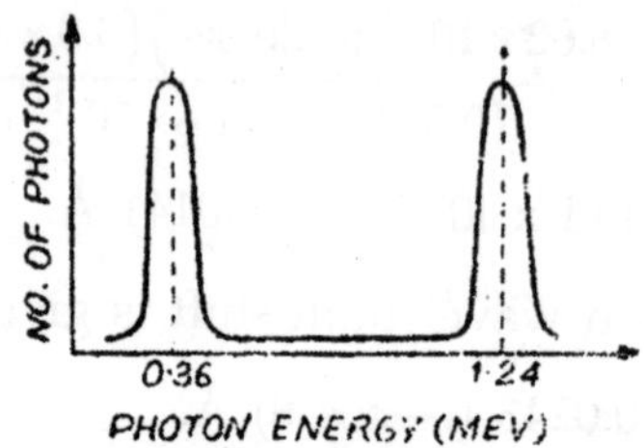

It can be worked out (as in above problems) that the wave-length of the 1.24-MeV incident photon is 0 01 Å, the Compton shift at 90° is 0.0243 Å, so that the modified wavelength is 0.0343Å. The energy corresponding to this wavelength is 0.36 MeV.

27. A photon collides with a free electron and is scattered at right angles to its initial direction. Let's find the percentage change in its energy as a result of this Compton collision, when the photon is a (i) microwave photon of λ = 3.0 cm, (ii) visible light photon of λ = 5000 Å (iii) X-ray photon of λ = 1.0 Å and (iv) y-ray photon of λ = 0.0124Å. The relative importance of the Compton effect in the different regions of electromagnetic spectrum is discussed below.

Solution: The change in photon energy is

$$\Delta E = hv - hv'$$

$$= \frac{hc}{\lambda} - \frac{hc}{\lambda'} = \frac{hc(\lambda' - \lambda)}{\lambda\lambda'} = \frac{hc\,\Delta\lambda}{\lambda\lambda'}$$

where $\Delta\lambda$ is the change (increase) in wavelength.

Thus

$$\Delta E = \frac{hc\,\Delta\lambda}{\lambda(\lambda+\Delta\lambda)} \qquad [\therefore \lambda' - \lambda = \Delta\lambda]$$

The initial energy of the photon is $E = hc/\lambda$. Therefore, the fractional change in energy is

$$\frac{\Delta E}{E} = \frac{\Delta\lambda}{\lambda+\Delta\lambda}$$

Now, the Compton shift is given by

$$\Delta\lambda = 0.0243\,(1-\cos\theta)\ \text{Å}$$
$$= 0.0243\text{Å for } \theta = 90°.$$

$$\therefore \qquad \frac{\Delta E}{E} = \frac{0.0243}{\lambda+0.0243}$$

Let us now calculate $\frac{\Delta E}{E}$ for the different photons:

(*i*) For microwave photon (λ = 3.0 cm = 3.0×10^8 Å); we have

$$\frac{\Delta E}{E} = \frac{0.0243}{(3.0\times10^8)+0.0243}$$
$$= 8\times10^{-11} = 8\times10^{-9}\ \%.$$

(*ii*) For visible light photon (λ = 5000Å), we have

$$\frac{\Delta E}{E} = \frac{0.0243}{5000+0.0243} = 4.8\times10^{-6} = 4.8\times10^{-4}\%$$

(*iii*) For the X-ray photon (λ = 1.0 Å), we have

$$\frac{\Delta E}{E} = \frac{0.0243}{1.0+0.0243} = 0.024 = 2.4\%$$

(*iv*) For the γ-ray photon (λ = 0.0124 Å), we have

$$\frac{\Delta E}{E} = \frac{0.0243}{0.0124+0.0243} = 0.66 = 66\%.$$

Thus the Compton effect is dominant only for the γ-ray region and the shorter X-ray region. It is not observable in the visible and microwave regions.

28. The change in wavelength of a photon for 90° scattering from a neutron is calculated as follow ($h = 6.6 \times 10^{-34}$ J-s, $c = 3 \times 10^{8}$ m/s, neutron mass = 1.0087 amu and 1 amu = 1.66×10^{-27} kg.)

Solution: The Compton change in wavelength is given by

$$\Delta\lambda = \frac{h}{m_0 c}(1 - \cos\theta)$$

where θ is the angle of scattering and m_0 is the rest mass of the scattering particle (here neutron). For $\theta = 90°$ we have

$$\Delta\lambda = \frac{h}{m_0 c}$$

$$= \frac{6.6 \times 10^{-34}\ \text{J - s}}{\left(1.0087 \times 1.66 \times 10^{-27}\ \text{kg}\right)\left(3 \times 10^{8}\ \text{m/s}\right)}$$

$$= 1.31 \times 10^{-15}\ \text{m}$$

$$= 1.31 \times 10^{-5}\ \text{Å}.$$

7

Crystal System

Space Lattice

Space Lattice in Crystal Structure: A crystal (such as rock-salt, quartz, calcite, diamond, mica, etc.) is a solid composed of *geometrically-regular* arrangement of atoms (or ions or molecules) in space. When struck into pieces, all pieces have the same shape with same angles between corresponding faces. These angles are characteristic of the given crystal.

Bravais assumed that a crystal is made up of a three-dimensional array of points such that each point is surrounded by the neighbouring points in an *identical* way. Such an array of points is known as 'Bravais lattice' or 'space lattice'. Thus, *a lattice is a regular arrangement of points extended repeatedly in space.*

Different Vectors of Translation

Mathematically, a lattice is expressed in terms of three translation vectors $\vec{a}, \vec{b}, \vec{c}$ such that the atomic arrangement looks exactly identical when viewed from points $\vec{r}$ and $\vec{r}'$ (Fig.), where

$$\vec{r}' = \vec{r} + u\vec{a} + v\vec{b} + w\vec{c} \cdot$$

u, v, w are arbitrary integers.

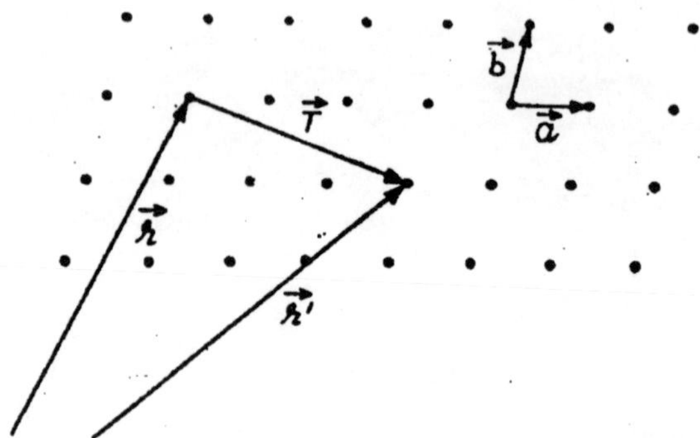

Basis: In the simplest crystals (such as copper, silver, sodium, etc) there is a single atom (or ion) at each lattice point. More often, however, there is a group of several atoms (or ions) attach lattice point. This group is called the 'basis'. Each basis is identical in composition, arrangement and orientation with any other basis.

The crystal structure is formed when a basis of atoms is attached identically to each lattice point, as shown in Fig.

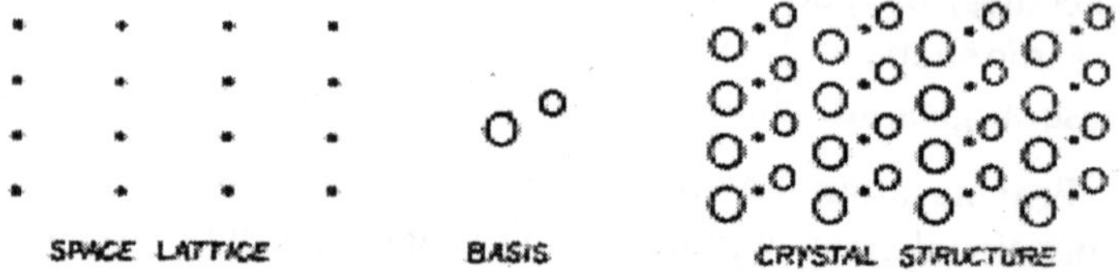

- Obviously, the logical relation is

  lattice + basis = crystal structure.

There is consequently a distinction between the structure of a crystal, which is the actual ordering in space of its constituent atoms, and the corresponding crystal lattice, which is a geometrical abstraction useful for classifying the crystal.

The translation vectors $\vec{a}, \vec{b}, \vec{c}$ of the lattice define the crystal axes.

A 'lattice translation operation' is defined as the displacement of a crystal parallel to itself by a crystal translation vector

$$\vec{T} = u\vec{a} + v\vec{b} + w\vec{c}$$

where the vector $\vec{T}$ connects any two lattice points (Fig.).

Unit Cell: The parallelepiped formed by the translation vectors $\vec{a}, \vec{b}, \vec{c}$ as edges is called a 'unit cell' of the space lattice. The angles between $(\vec{b}, \vec{c})$, $(\vec{c}, \vec{a})$ and $(\vec{a}, \vec{b})$ are denoted by α, β and γ respectively (Fig.).

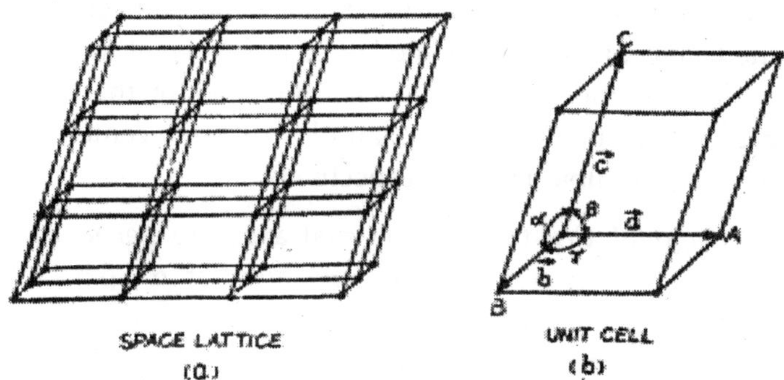

Fig. shows part of a space lattice. It is possible to isolate a unit cell, and the space lattice can be constructed by repeatedly translating the unit cell along its edges.

Systems of Crystal

There are seven types of crystals depending upon their axial ratios ($a : b : c$) and angles between them (α, β, γ). Bravais showed that they can give rise to 14 types of space lattices. These crystals and the corresponding Bravais lattices with their characteristic features are as below:

Cubic Crystals: In cubic crystals, the crystal axes are perpendicular to one another ($\alpha = \beta = \gamma = 90°$) and the repetitive interval is the same along the three axes ($a = b = c$). Cubic lattices may be simple, body-centered or face-centered as shown in below Fig.

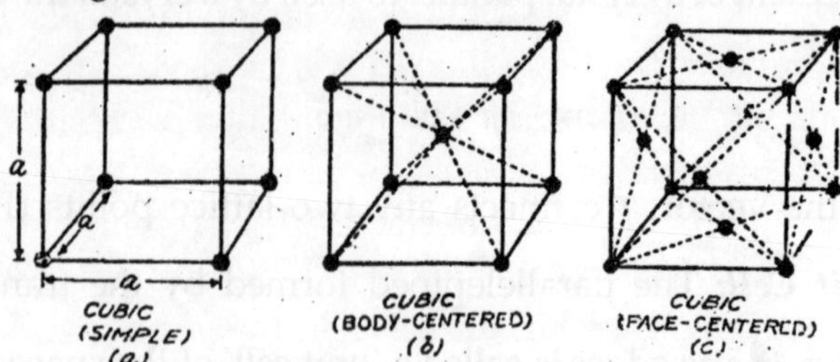

In the simple cubic lattice, the lattice points are situated only at the corners of the unit cell (Fig *a*). In the body-centered lattice the lattice points are situated at the corners and also at the intersection of the body diagonals of the unit cell (Fig. *b*). In the face-centered lattice, the lattice points lie at the corners as well as at the centers of all the six faces of the unit cell (Fig. *c*). CsCl and NaCl are examples of body-centered and face-centered cubic lattices respectively.

Tetragonal Crystals: The crystal axes are perpendicular Solid State Physics and Devices to one another ($\alpha = \beta = \gamma = 90°$). The repetitive intervals along two axes are the same, but the interval along the third axis is different ($a = b \neq c$). Tetragonal lattices may be simple or body-centered.

Orthorhombic Crystals: The crystal axes are perpendicular to one another ($\alpha = \beta = \gamma = 90°$), but the repetitive intervals are different along all three axes ($a \neq b \neq c$). Orthorhombic crystals may be simple, base-centered, body-centered, or face-centered.

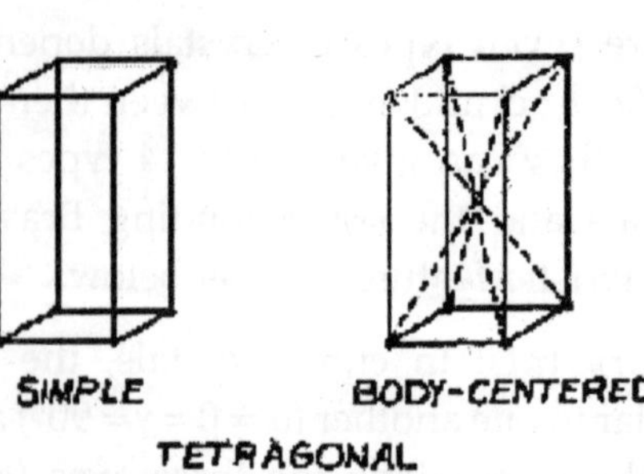

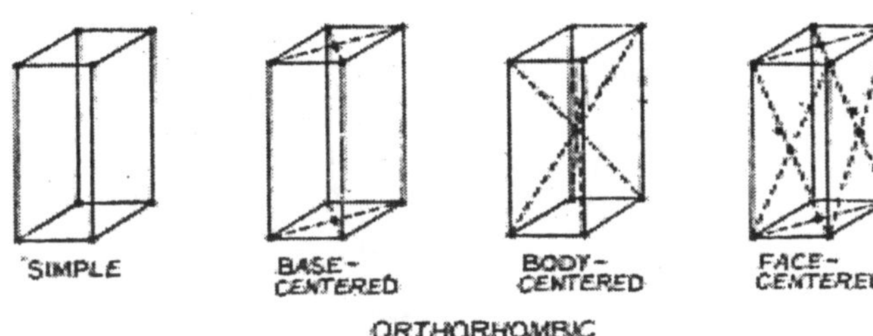

ORTHORHOMBIC

Monoclinic Crystals: Two of the crystal axes are not perpendicular to each other, but the third is perpendicular to both of them ($\alpha = \gamma = 90° \neq \beta$). The repetitive intervals are different along all three axes ($a \neq b \neq c$). Monoclinic lattices may be simple or base-centered (Fig.).

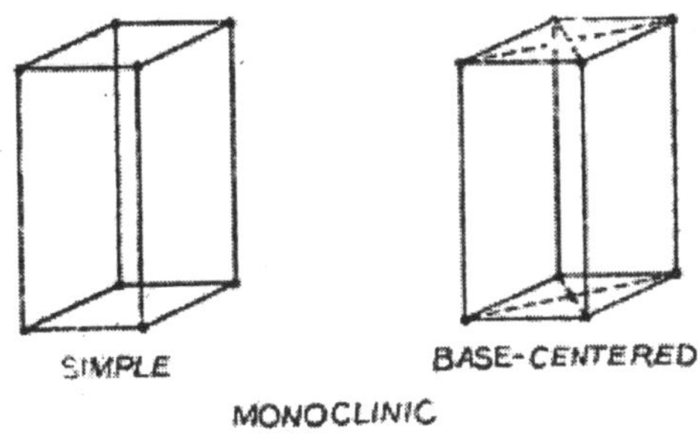

MONOCLINIC

Triclinic Crystals: None of the crystal axes is perpendicular to any of the others ($\alpha \neq \beta \neq \gamma$), and the repetitive intervals are different along all three axes ($a \neq b \neq c$). The triclinic lattice is only simple.

Trigonal (or Rhombohedral) Crystals: The angles between each pair of crystal axes are the same but different from 90° ($\alpha = \beta = \gamma \neq 90°$). The repetitive interval is the same along all three axes ($a = b = c$). The trigonal lattice is only simple (Fig.).

Hexagonal Crystals: Two of the crystal axes are 60° apart while the third is perpendicular to both of them ($\alpha = \beta = 90°$, $\gamma = 120°$). The repetitive intervals are the same along the two 60°-apart axes, but the interval along the third is different ($a = b \neq c$). The hexagonal lattice is only simple (Fig.).

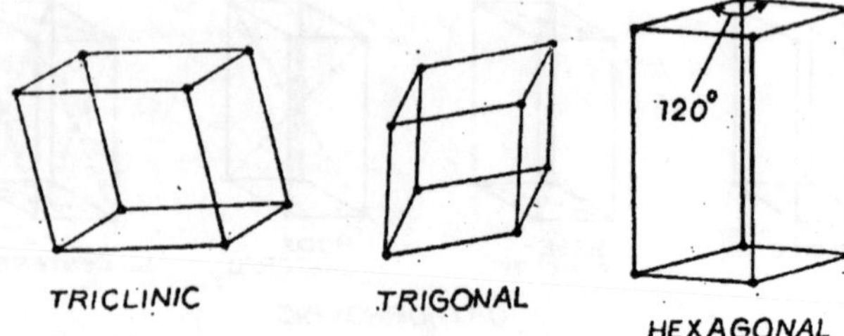

The above description can be tabulated as below:

Crystal class	*Intercepts*	*Angles between on Axes*	*Types of Bravias Axes space lattices*
Cubic	$a = b = c$	$\alpha = \beta = \gamma = 90°$	simple, body-centered, face-centered
Tetragonal	$a = b \neq c$	$\alpha = \beta = \gamma = 90°$	simple, body-centered
Orthorhombic	$a \neq b \neq c$	$\alpha = \beta = \gamma = 90°$	simple, base-centered, body-centered, face centered
Monoclinic	$a \neq b \neq c$	$\alpha = \gamma = 90° \neq \beta$	simple, base-centered
Triclinic	$a \neq b \neq c$	$\alpha \neq \beta \neq \gamma$	simple
Trigonal	$a = b =c$	$\alpha = \beta = \gamma \neq 90°$	simple
Hexagonal	$a = b \neq c$	$\alpha = \beta = 90°, \gamma = 120°$	simple

Cubic Crystal Lattices: The type of crystal lattice that is equivalent in all directions in space is the "cubic" lattice ($a = b = c; \alpha = \beta = \gamma = 90°$). There are three types of cubic lattices; simple, body-centered and face-centered. Let us consider them in detail:

Simple Cubic (sc) Lattice: In this space lattice, the lattice points are situated only at the corners of the unit cells constituting the three-dimensional structure (Fig.). Each cell has eight corners, and eight cells meet at each corner. Thus only one-eight of a lattice point belongs to each cell. That is, *there is only 1 lattice point (or 1 atom) per unit cell.* A unit cell containing only 1 lattice point is called a 'primitive' cell. Since the simple cubic lattice is built of primitive cells, it is also known as cubic *P* lattice.

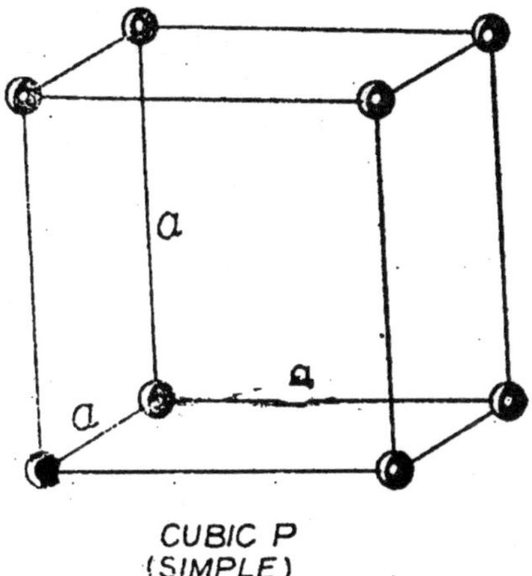

CUBIC P
(SIMPLE)

Coordination Number: *The coordination number is defined as the number of nearest neighbours around any lattice point (or atom) in the crystal lattice.* Let us compute it for the simple cubic lattice (Fig.).

Taking any one of the lattice points as origin, and the three edges passing through that point as x, y and z axes, the positions of the *nearest* neighbours of the origin are

$$\pm a\hat{i}, \pm a\hat{j}, \pm a\hat{k}$$

where $\hat{i}$, $\hat{j}$, $\hat{k}$ are unit vectors along the x, y, z axis respectively. The coordinates of these points (or atoms) nearest the origin are

$$(\pm a, 0, 0); (0, \pm a, 0); (0, 0, \pm a).$$

Their number is obviously 6. Hence *the coordination number of a simple cubic lattice is 6.* The distance between two nearest neighbours is a, which is the 'lattice constant' (length of each edge of the unit cell).

Body-centered Cubic (bcc) Lattice: In this lattice, the lattice points (or atoms) are situated at each corner of the unit cell and also at the Intersection of the *body* diagonals of the cell (Fig.). This lattice is also known as cubic I lattice.

Each cell has 8 corners and 8 cells meet at each corner. Thus only 1/8th of a lattice point (or atom) at a corner belongs to any one cell. Also there is a lattice point (or atom) at the body centre of each cell. Thus the total number of lattice points in any one cell is $\left(\frac{1}{8}\times 8\right)$ + 1=2, that is, *a bcc lattice has 2 lattice points (or atoms) per unit cell.*

To compute the coordination number, let us take the lattice point at the body centre as the origin; the x, y, z axes being parallel to the edges of the unit cell. The positions of the *nearest* neighbours of the origin are

$$\left(\pm\frac{a}{2}\hat{i}, \pm\frac{a}{2}\hat{j}, \pm\frac{a}{2}\hat{k}\right)$$

The coordinates of these points (or atoms) nearest the origin are

$$\left(\frac{a}{2},\frac{a}{2},\frac{a}{2}\right);\left(-\frac{a}{2},\frac{a}{2},\frac{a}{2}\right);\left(\frac{a}{2},-\frac{a}{2},\frac{a}{2}\right);$$

$$\left(\frac{a}{2},\frac{a}{2},-\frac{a}{2}\right);\left(-\frac{a}{2},-\frac{a}{2},\frac{a}{2}\right);$$

$$\left(-\frac{a}{2},\frac{a}{2},-\frac{a}{2}\right);\left(\frac{a}{2},-\frac{a}{2},-\frac{a}{2}\right);$$

$$\left(-\frac{a}{2},-\frac{a}{2},-\frac{a}{2}\right).$$

Their number is obviously 8. Hence *the coordination number of a bcc lattice is 8.*

The distance between any two nearest neighbours is

$$\sqrt{\left[\left(\frac{a}{2}\right)^2+\left(\frac{a}{2}\right)^2+\left(\frac{a}{2}\right)^2\right]} = \frac{\sqrt{3}}{2}a$$

A typical bcc structure is CsCl, and so the bcc lattice is often known as the CsCl structure. In this structure each ion is surrounded by 8 neighbours of the opposite charge (Fig.).

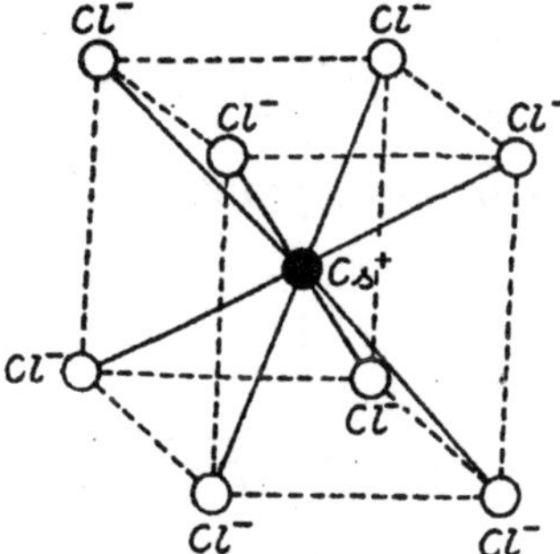

Face-centered Cubic (fcc) Lattice: In this lattice, the lattice points (or atoms) are situated at all the eight corners of the unit cell and also at the centers of all the six faces of the cell. This lattice is also known as cubic *F* lattice.

Each unit cell has 8 corners and 8 cells meet at each corner. Thus only 1/8th of a lattice point (or atom) at a corner belongs to any one cell. Similarly, a lattice point at the center of a face of the cell is shared by 2 cells, that is, only ½ of lattice point belongs to any one cell. Since a cell has 8 corners and 6 faces, the total number of lattice points belonging to any one cell is $\left(\frac{1}{8}\times 8\right)+\left(\frac{1}{2}\times 6\right)$ = 4, that is, *an fcc lattice has 4 lattice points (or atoms) per unit cell.*

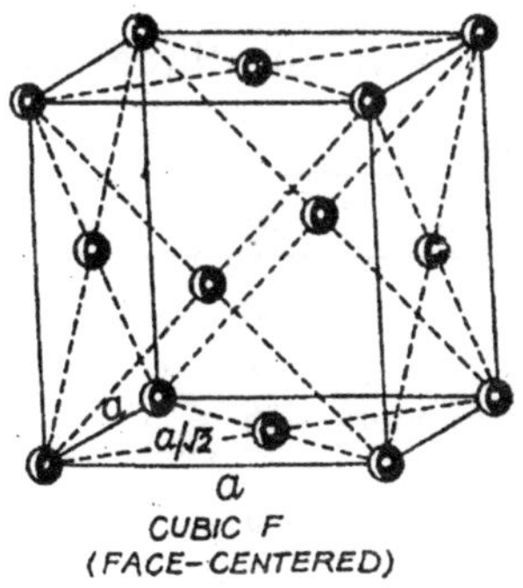

CUBIC F
(FACE–CENTERED)

To compute the coordination number, let us take any of the lattice points as the origin; the x, y, z axes being parallel to the edges of the unit cell. The positions of the *nearest* neighbours of the origin are

$$\left(\pm\frac{a}{2}\hat{i}\pm\frac{a}{2}\hat{j}, \pm\frac{a}{2}\hat{j}\pm\frac{a}{2}\hat{k}, \pm\frac{a}{2}\hat{k}\pm\frac{a}{2}\hat{i}\right)$$

The coordinates of these points (or atoms) are

$$\left(\frac{a}{2},\frac{a}{2},0\right);\left(-\frac{a}{2},\frac{a}{2},0\right);\left(\frac{a}{2},-\frac{a}{2},0\right);$$

$$\left(-\frac{a}{2},-\frac{a}{2},0\right);\left(0,\frac{a}{2},\frac{a}{2}\right);\left(0,-\frac{a}{2},\frac{a}{2}\right);$$

$$\left(0,\frac{a}{2},-\frac{a}{2}\right);\left(0,-\frac{a}{2},-\frac{a}{2}\right);\left(\frac{a}{2},0,\frac{a}{2}\right);$$

$$\left(-\frac{a}{2},0,\frac{a}{2}\right);\left(\frac{a}{2},0,-\frac{a}{2}\right);\left(-\frac{a}{2},0,-\frac{a}{2}\right).$$

There number is obviously 12. Hence *the coordination number of an fcc lattice is 12.*

The distance between two nearest neighbours is

$$\sqrt{\left[\left(\frac{a}{2}\right)^2+\left(\frac{a}{2}\right)^2+(0)^2\right]}=\frac{1}{\sqrt{2}}a.$$

This lattice gives a very efficient packing (more atoms per unit volume) and so it is usually the most stable structure. The most familiar fcc lattice is NaCl, and so the fcc lattice is often called NaCl structure.

The characteristics of the three cubic lattices may be tabulated as below:

	Simple	***Body-centered***	***Face-centered***
Volume of unit cell	a^3	a^3	a^3
Lattice points per cell	1	2	4
Number of nearest neighbours	6	8	12
Nearest-neighbour distance	a	$\frac{\sqrt{3}}{2}a$	$\frac{1}{\sqrt{2}}a$

NaCl Structure: The NaCl crystal is a system of Na^+ and Cl^- ions arranged alternately in a 'cubic' pattern in space so that the electrostatic attraction between the nearest neighbours is maximum. Each ion lies on three rows of equally-spaced ions at right angles to one another.

In Fig. is shown a unit cell of NaCl lattice. The Na^+ ions are situated at the corners as well as at the centres of the faces of the cube, that is, Na^+ ions lie on a fcc lattice. So do the Cl^- ions, their lattice being relatively displaced half the edge of the unit cell along each axis. Thus NaCl crystal can be thought of as composed of interleaved fcc Na^+ and Cl^- sublattices.

Each cell has 8 corners and 8 cells meet at each corner. Thus an ion at a corner of the cell is shared by 8 cells, *i.e.*, only $\frac{1}{8}$ ion belongs to any one cell. Similarly, an ion at the centre of a face of the cell is shared by 2 cells, *i.e.* only ½ ion belongs to any one cell. Since a cell has 8 corners and 6 faces, it has $\left(\frac{1}{8} \times 8\right) + \left(\frac{1}{2} \times 6\right) = 4$ ions of one kind, and similarly 4 ions of the other kind. Thus *there are* 4 $Na^+ - Cl^-$ *ion pairs per unit cell.*

Each Na^+ ion has 6 Cl^- ions as nearest neighbours, and similarly each Cl^- ion has 6 Na^+ ions. Hence the coordination number of NaCl is 6, the same as that for simple cubic lattice.

Since a given Na^+ ion is attracted by 6 nearest Cl^- neighbours, and does not belong to any single Cl^- ion, the NaCl crystal cannot be thought of as being composed of molecules.

Unit Cell and Lattice Constant of a Space Lattice: A unit cell is a parallelopiped isolated from a space lattice, having its edges along the translation vectors. The space lattice can be constructed by repeatedly translating the unit cell along its edges.

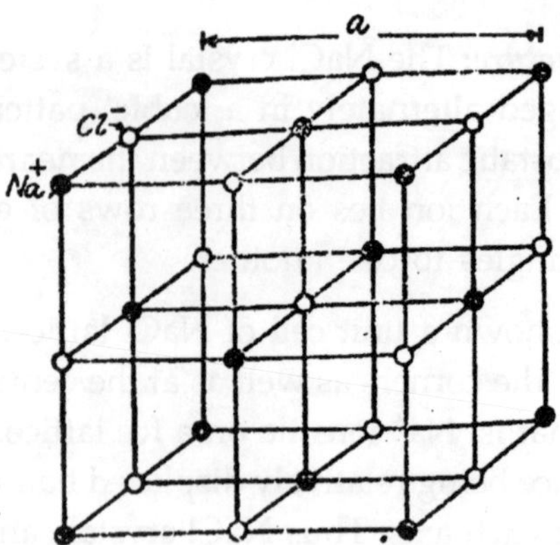

The lattice constant of a cubic lattice is the dimension of its unit cell. In the figure above, the lattice constant is *a*.

Let us consider a cubic crystal lattice of lattice constant a. The volume of the unit cell is a^3. If ρ be the density of the cell, then

$$\rho = \frac{m}{a^3} \qquad ...(i)$$

where *m* is the mass per unit cell.

Let *n* be the number of molecules per unit cell. The mass of one molecule is *M*/*N*, where *M* is the molecular weight of the crystal and *N* is Avogadro's number. Then, the mass per unit cell is

$$m = n\frac{M}{N}$$

Substituting this value of *m* in eq. (*i*), we get

$$\rho = \frac{nM}{Na^3}$$

or

$$a = \left(\frac{nM}{N\rho}\right)^{\frac{1}{3}}.$$

From this relation the value of the lattice constant can be calculated.

Lattice Planes of a Crystal: A crystal lattice may be considered as an aggregate of a set of parallel, equally-spaced planes passing through the lattice points. The planes are called 'lattice planes', and the perpendicular distance between adjacent planes is called 'interplanar spacing'.

A given space lattice may have an infinite sets of lattice planes, each having its characteristic interplanar spacing. In below given figure are shown three sets of lattice planes with interplanar spacings d_1, d_2 and d_3. Each set accommodates *all* the lattice points. Out of these, only those which have high density of lattice points are significant and show diffraction of X-rays. They are known as 'Bragg planes' or 'Cleavage planes'. When a crystal is struck, it breaks most easily across its cleanvage planes.

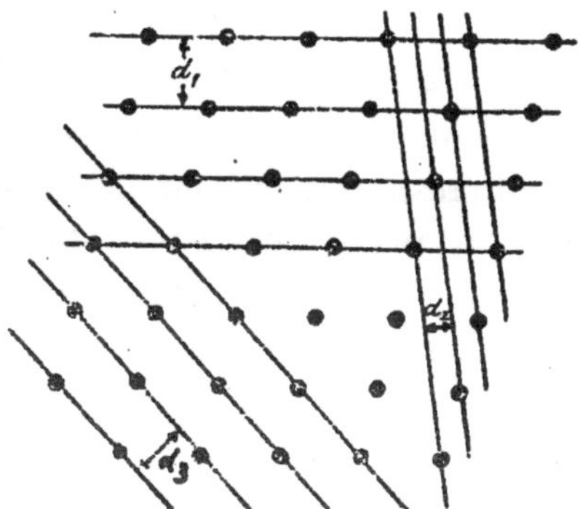

Position and Orientation of Lattice Planes in a Crystal— Miller Indices: *The position and orientation of a lattice plane in a crystal are determined by three smallest whole numbers which have same ratios with one another as the reciprocals of the intercepts of the plane on the three crystal axes.* These numbers, denoted by *h, k, l,* are known as the 'Miller indices' of that plane or of any plane parallel to it); and the plane is specified as (*hkl*).

Let *OX, OY, OZ* be three axes parallel to the crystal axes the planes *XOY, XOZ* and *ZOY* are parallel to the faces of the crystal. Besides these, other possible cleavage planes also exist.

Let ABC be a standard plane cutting all the three axes at A, B and C with intercepts OA, OB and $OC = a, b$ and c respectively, where $\vec{a}, \vec{b}$ and $\vec{c}$ are the translation vectors of the crystal lattice.

The directions of the other faces of the crystal are governed by the 'law of rational indices'. This law states that *a face which is parallel to a plane whose intercepts on the three axes are m_1a, m_2b and m_3c where m_1, m_2 and m_3 are small whole numbers, is a possible face of the crystal.* Thus, if $A'B'C'$ be a possible crystal face, then

$$OA' : OB' : OC' = m_1a : m_2b : m_3c$$

$$= \frac{a}{m_2m_3} : \frac{b}{m_1m_3} : \frac{c}{m_1m_2}$$

$$= \frac{a}{h} : \frac{b}{k} : \frac{c}{l}$$

where h $(= m_2 m_3)$, k $(= m_1m_3)$ and $l(= m_1m_2)$ are again small whole numbers. The numbers h, k, l are the Miller indices of the plane $A'B'C'$ with respect to the standard plane. The Miller indices of the standard plane are always (111). If a plane cuts an axis on the negative side of the origin, the corresponding index is negative, indicated by placing a minus sign above the index, as $(h, \bar{k}, l)$.

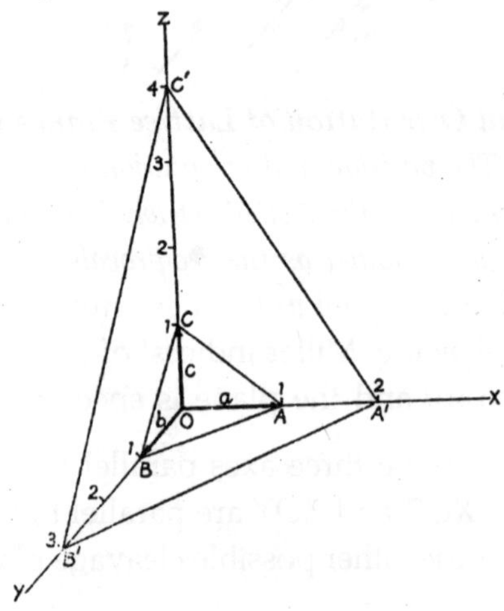

As an example, suppose a plane cuts the X-axis at $2a$, the Y-axis at $3b$ and the Z-axis at $4c$. Then, from the law of rational indices, we have

$$2a : 3b : 4c = \frac{a}{h} : \frac{b}{k} : \frac{c}{l}$$

or $$h : k : l = \frac{1}{2} : \frac{1}{3} : \frac{1}{4}$$

The smallest whole numbers which have same ratios as the reciprocals $\frac{1}{2}, \frac{1}{3}$ and $\frac{1}{4}$ are 6 : 4 : 3. Hence the miller indices of the plane are $h = 6$, $k = 4$ and $l = 3$; or the plane is (643).

Directional Indices: The indices of a direction in a crystal are the set of the smallest integers which have the same ratios as the components of a vector in the desired direction, referred to the axes. The directional indices (integers) are written as $[uvw]$. Thus the X-axis is the [100] direction, the —Y-axis is the $\left[01\bar{0}\right]$ direction, and so on.

In cubic crystals, the normal to a plane (hkl) is the direction $[hkl]$.

Sketching of Lattice Planes in a Cubic Crystal: Let us consider a simple cubic lattice plane ($\alpha = \beta = \gamma = 90°$ and $a = b = c$) The axes OX, OY, OZ form a right-angled set. The face $ABFE$ in Fig (a), or any plane parallel to it, has an intercept on X-axis, but it is parallel to the Y and Z-axis, *i.e.* the intercepts on the Y and Z-axis are each infinity.

If the intercept on the X-axis is taken as 1, then the miller indices are the reciprocals of 1, ∞, ∞, *i.e.*, 1, 0, 0; and the plane is (100). The plane parallel to $ABFE$ but intercepting the X-axis on the negative side of the origin is $\left(\bar{1}00\right)$.

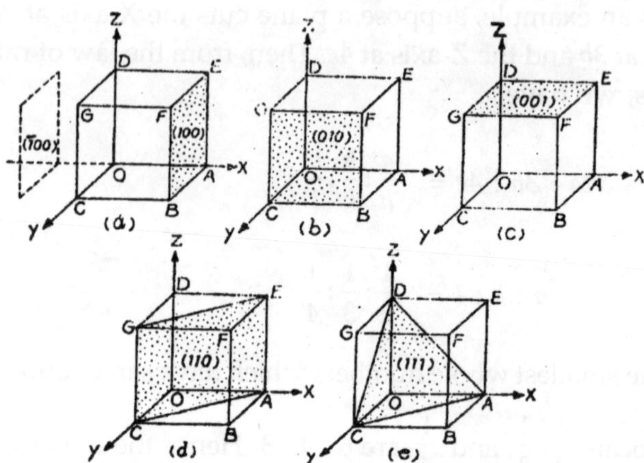

Similarly a plane parallel to *BCGF* is a (010) plane (Fig. *b*) and the one parallel to *DEFG* is a (001) plane (Fig. *c*).

The diagonal plane *ACGE* (Fig. *d*) has equal intercepts of 1 on *X* and *Y* axes and is parallel to the *Z*-axis. It is therefore a (110) plane.

The plane *ACD* (Fig. *e*) makes equal intercepts on the three axes and so its miller indices are 1, 1, 1 and so it is a (111) plane.

Interplanar Spacing: The separation between successive lattice planes of cubic, tetragonal and orthorhombic crystals, for which $\alpha = \beta = \gamma = 90°$, can be deduced as follows.

Let *OX, OY* and *OZ* be three axes parallel to the crystal axes (Fig.). Let *ABC* be one of a series of parallel lattice planes in the crystal. Let the plane *ABC* have intercepts $OA = a/h$, $OB = b/k$ and $OC = c/l$. Let us suppose that the origin *O* lies in the plane *adjacent* to *ABC*. Then, *ON*, the length of the normal from the origin to the plane *ABC*, is equal to the interplanar distance*d*. Let θ_a, θ_b and θ_c be the angles which *ON* makes with the three crystallographic axes respectively. Then the direction-cosines of *ON* are

$$\cos\theta_a = \frac{ON}{OA},\ \cos\theta_b = \frac{ON}{OB} \text{ and } \cos\theta_c = \frac{ON}{OC}.$$

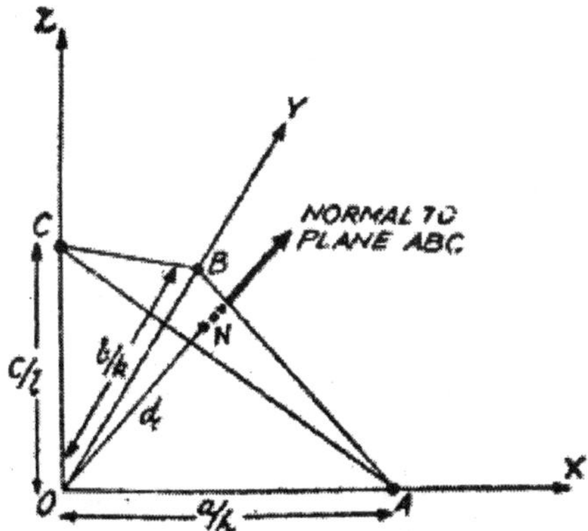

We know that the sum of the squares of the direction-cosines of a line is equal to unity ($\cos^2\theta_a + \cos^2\theta_b + \cos^2\theta_c = 1$). Therefore

$$\left(\frac{ON}{OA}\right)^2 + \left(\frac{ON}{OB}\right)^2 + \left(\frac{ON}{OC}\right)^2 = 1$$

or $$\left(\frac{d}{a/h}\right)^2 + \left(\frac{d}{b/k}\right)^2 + \left(\frac{d}{c/l}\right)^2 = 1$$

or $$d^2\left[\frac{h^2}{a^2} + \frac{k^2}{b^2} + \frac{l^2}{c^2}\right] = 1$$

or $$d = \frac{1}{\sqrt{\left(\frac{h^2}{a^2} + \frac{k^2}{b^2} + \frac{l^2}{c^2}\right)}}$$

For a cubic crystal, the lengths of the sides of a unit cell are equal, *i.e.* $a = b = c$.

$\therefore$ $$\boxed{d = \frac{a}{\sqrt{\left(h^2 + k^2 + l^2\right)}}}$$

Let us calculate the values of d for some of the series of lattice planes in a *simple* cubic crystal. For the (100) planes, we have $h = 1$, $k = 0$, and $l = 0$

$\therefore \qquad d_{100} = a.$

For the (110) planes, we have $h = 1$, $k = 1$, and $l = 0$

$\therefore \qquad d_{110} = a\sqrt{2}\cdot$

For the (111) planes, we have $h = 1$, $k = 1$, and $l = 1$.

$\therefore \qquad d_{111} = a\sqrt{3}\cdot$

Thus$d_{100} : d_{110} : d_{111} = 1 : \frac{1}{\sqrt{2}} : \frac{1}{\sqrt{3}} = 1 : 0.71 : 0.58.$

Spacing between Lattice Planes in BCC and FCC Crystals: Fig. (*a*) and (*b*) show bcc lattices. Comparing with simple cubic lattice, it is seen that in bcc lattice there exist additional planes half-way between the (100) planes and also between the (111) planes.

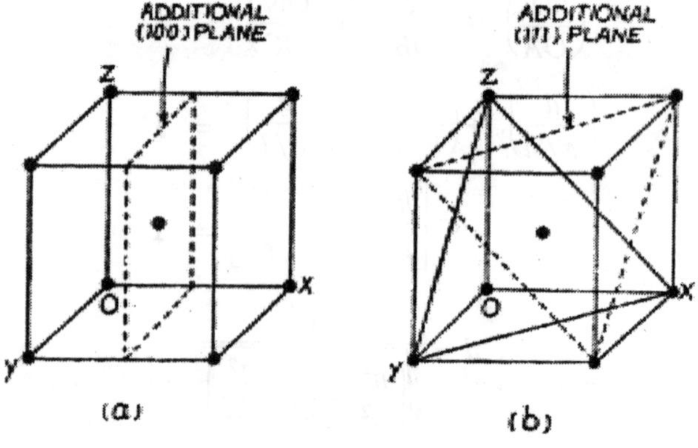

(a) (b)

Therefore, in the light of spacings in the simple cubic lattice ($d_{100}=a, d_{110}= a/\sqrt{2}, d_{111}= a/\sqrt{3}$), the spacings between (100), (110) and (111) planes in the bcc lattice are given by

$$d_{100} = \frac{a}{2}$$

$$d_{110} = \frac{a}{\sqrt{2}}$$

and $$d_{111} = \frac{a}{2\sqrt{3}}.$$

Thus $$d_{100} : d_{110} : d_{111} = 1 : \sqrt{2} : \frac{1}{\sqrt{3}}$$

Fig. (*a*) and (*b*) show fcc lattices. Again comparing with simple cubic lattice, in fee lattice there exist additional planes halfway between the (100) planes and also between the (110) planes.

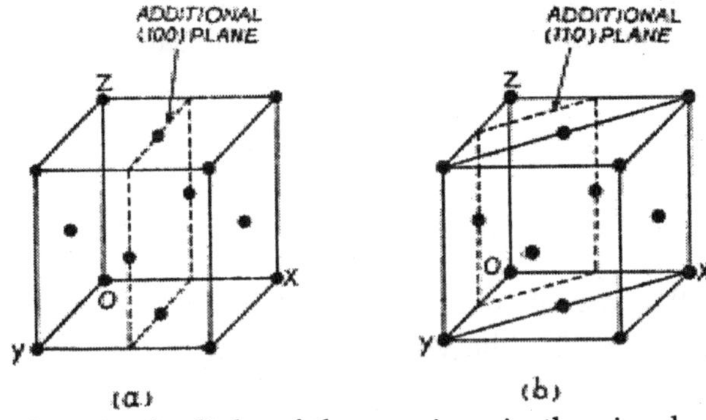

Therefore, in the light of the spacings in the simple cubic lattice, the spacings between (100), (110) and (111) planes in the fee lattice are given by

$$d_{100} = \frac{a}{2}$$

$$d_{110} = \frac{a}{2\sqrt{2}}$$

and $$d_{111} = \frac{a}{\sqrt{3}}$$

Thus, $$d_{100} : d_{110} : d_{111} = 1 : \frac{1}{\sqrt{2}} : \frac{2}{\sqrt{3}}.$$

Density of Lattice Points in a Lattice Plane: Let us consider *N* parallel lattice planes in a crystal lattice. Let *A* be the cross-sectional area of each lattice plane and *d* the spacing between successive planes. The volume of the lattice space under consideration is then NAd. If V be the volume of a unit cell, then the number of unit cells in the lattice space is $\frac{NAd}{V}$.

If n be the number of lattice points per unit cell, then the total number of lattice points in the lattice space is $\frac{nNAd}{V}$.

Now, let ρ be the density of lattice points in a lattice plane, *i.e.*, the number of lattice points per unit area of the plane. Then the total number of lattice points in the same lattice space is NAP. Thus,

$$NA\rho = \frac{nNAd}{V}$$

or

$$\rho = \frac{nd}{V}$$

In the case of cubic, tetragonal and orthorhombic lattices ($\alpha = \beta = \gamma = 90°$), the volume of a unit cell is $V = abc$. Thus

$$\rho = \frac{nd}{abc}.$$

For primitive lattice in each of these systems, there is one lattice point per unit cell, *i.e.*, $n = 1$. In such cases

$$\rho = \frac{d}{abc}$$

Hexagonal Close-packed Structure—(HCP): A crystal whose constituents atoms are so arranged as to occupy the least possible volume is said to have a 'close-packed' structure. Such structures occur when the bonding forces are spherically symmetric (as in inert gases) or very nearly so (as in metals).

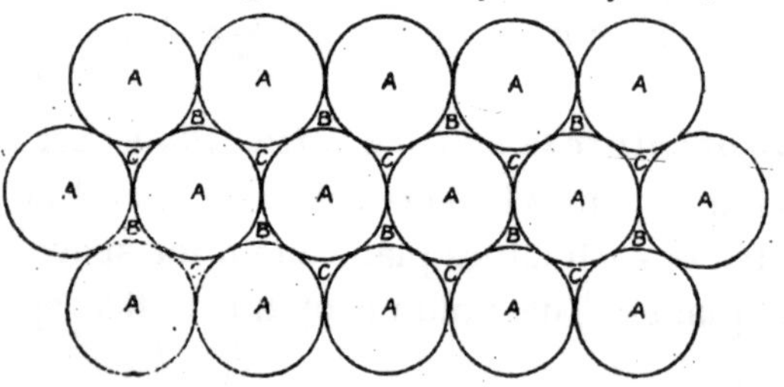

To understand the close-packed structure, let us consider identical spheres. The spheres can be arranged in a single *closest-packed* layer by placing each sphere in contact with six others (Fig.). The layer then assumes hexagonal shape. A second identical layer can be formed over this by placing spheres on the hollows *B*, each formed by three spheres in the bottom layer. Now, a third layer can be added in two different ways. In one, the spheres in the third layer are placed over the hollows *C;* the resulting three-dimensional structure is then 'face-centered cubic' (or 'cubic close-packed') structure. Thus the packing in the fcc structure is *ABC ABC ABC*.........

Alternatively, the spheres in the third layer can be placed over *A*'s, that is, directly over the spheres in the first layer; the resulting three-dimensional structure is then 'hexagonal closed-packed' (hcp). Thus the packing in the hcp structure is *AB AB AB*.........

In the hcp structure, the spheres in *alternate* layers are directly above one another (Fig.). The atom positions in this structure do not constitute a space lattice. The space lattice is simple hexagonal with a basis of two identical atoms associated with each lattice point.

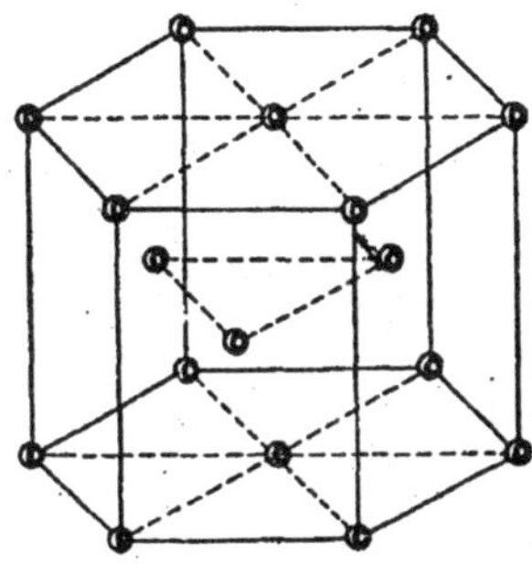

The coordination number in a closed-packed crystal (either fcc or hep) is 12. In both cases each sphere in a particular layer fits into the hollow formed by three spheres in the layer below it; hence each sphere is in contact with three spheres in the layer below it, and with three in the layer above it as well. Adding these six spheres to the six it touches in its own layer, each sphere in a close-packed structure has 12 nearest neighbours,,; that is, the coordination number is 12.

Beryllium, magnesium, cobalt and zinc are among the elements with hcp structures.

In a face-centered cubic (fcc) structure, also called as cubic close-packed, the sequence repeats itself every three layers, instead of every two layers. Copper, lead, gold and argon are examples of close-packed fcc structures.

Formation of Diamond

The space lattice of diamond is face-centered cubic (fcc) with a basis of two carbon atoms associated with each lattice point. The figure below shows the positions of atoms in the cubic cell of the diamond structure projected on a cubic face. The fractions denote height above the base in units of a cube edge. The points at 0 and ½ are on the fcc lattice, those at ¼ and ¾ are on a similar lattice displaced along the body diagonal by one-fourth of its length. Thus the diamond lattice is composed of two inter-leaved fcc sublattices, one of which is shifted relative to the other by one-fourth of a body diagonal.

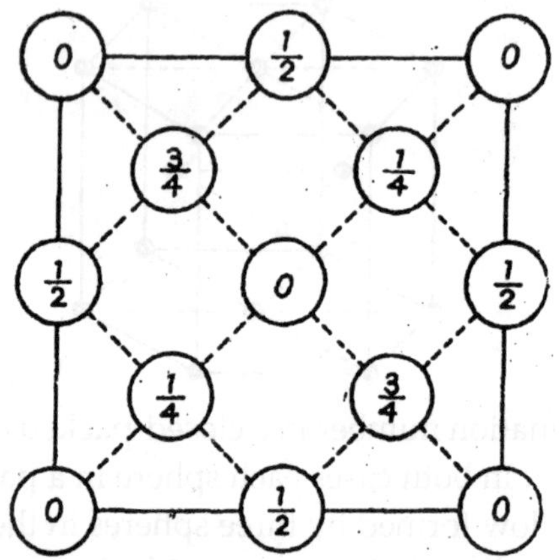

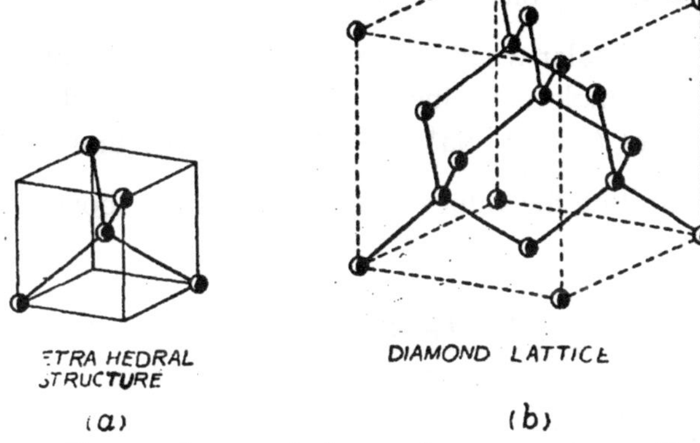

(a) (b)

In a diamond crystal the carbon atoms are linked by directional covalent bonds. Each carbon atom forms covalent bonds with four other carbon atoms that occupy four corners of a cube in a tetrahedral structure (Fig. a). The length of each bond is 1.54 Å and the angle between the bonds is 109.5°. The entire diamond lattice is constructed of such tetrahedral units (Fig. b).

In the diamond lattice, each atom has four nearest neighbours with which it forms covalent bonds. Thus the coordination number of diamond crystal is 4. The number of atoms per unit cell is 8.

Silicon, germanium and gray tin crystallise in the diamond structure.

PROBLEMS

1. ***The lattice constant of NaCl crystal is calculated. The density of NaCl is 2189 kg/m^3 and Avogadro's number N is 6.02 × 10^{26}/kg-molecule.***

Solution: NaCl is a cubic crystal having an fcc lattice. The lattice constant of a cubic lattice is given by

$$a = \left(\frac{nM}{N\rho}\right)^{\frac{1}{3}},$$

where M is molecular weight and ρ is the density of the crystal. n is number of molecules per unit cell.

The molecular weight of NaCl is 58.5 kg/kg-molecule (or 58.5 gm/gm-molecule). Since it belongs to fcc lattice, the number of molecules (in fact, $Na^+ - Cl^-$ ion pairs) per unit cell is 4.

Substituting the values of n, M, N, ρ in the above relation, we have

$$a = \left[\frac{4 \times 58.5 \text{ kg}/(\text{kg - molecule})}{6.02 \times 10^{26}/(\text{kg - molecule}) \times 2189 \text{ kg}/\text{m}^3}\right]^{\frac{1}{3}}$$

$$= (177 \times 10^{-30} \text{ m}^3)^{1/3}$$

$$= 5.61 \times 10^{-10} \text{ m} = 5.61 \text{ Å}.$$

2. *The lattice constant of KBr from the following data is calculated as follows: density = 2.7 gm/cm³, molecular weight = 119, Avogadro's number = 6.02 × 10²³/gm-molecule. KBr is fcc lattice.*

Solution: The lattice constant for a cubic lattice is given by

$$a = \left(\frac{nM}{N\rho}\right)^{\frac{1}{3}}$$

where n is number of lattice points (atoms or molecules) per unit cell. For fcc lattice, $n = 4$.

$$\therefore\ a = \left[\frac{4 \times 119 \text{ gm}/(\text{gm - molecule})}{6.02 \times 10^{23}/(\text{gm - molecule}) \times 2.7 \text{ gm}/\text{cm}^3}\right]^{\frac{1}{3}}$$

$$= [293 \times 10^{-24} \text{ cm}^3]^{1/8}$$

$$= 6.64 \times 10^{-8} \text{ cm} = 6.64 \text{ Å}.$$

3. *Copper has a density of 8.96 gm/cm³ and an atomic weight of 63.5. The distance between two nearest copper atoms in the fcc structure is calculated. The Avogadro number is 6.02 × 10²³.*

Solution: The lattice constant a for a cubic lattice is given by

$$a = \left(\frac{nM}{N\rho}\right)^{\frac{1}{3}},$$

where n is number of atoms per unit cell, M is atomic weight and ρ is density. For fcc structure, $n = 4$.

$$\therefore\ a = \left[\frac{4 \times 63.5\ \text{gm}/(\text{gm-atom})}{6.02 \times 10^{23}/(\text{gm-atom}) \times 8.96\ \text{gm}/\text{cm}^3}\right]^{\frac{1}{3}}$$

$$= [47.1 \times 10^{-24}\ \text{cm}^3]^{1/3}$$

$$= 3.61 \times 10^{-8}\ \text{cm} = 3.61\ \text{Å}.$$

The nearest-neighbour distance in fcc lattice is $a/\sqrt{2}$. Therefore, the distance between two nearest copper atoms

$$= \frac{3.61\ \text{Å}}{\sqrt{2}}$$

$$= 2.55\ \text{Å}.$$

4. ***The density of α-iron is 7870 kg/m³ and its atomic weight is 55.8. Given that α-iron crystallises in bcc space lattice, its lattice constant is deduced as follows. Avogadro's number N = 6.02 × 10²⁶ per kg-atom.***

Solution: The lattice constant for a cubic crystal of atomic weight M and density ρ is given by

$$a = \left(\frac{nM}{N\rho}\right)^{\frac{1}{3}},$$

where n is the number of atoms per unit cell. For bcc lattice, $n = 2$.

$$\therefore a = \left[\frac{2 \times 55.8\ \text{kg/(kg-atom)}}{6.02 \times 10^{26}/\text{(kg-atom)} \times 7870\ \text{kg/cm}^3}\right]^{\frac{1}{3}}$$

$$= [23.5 \times 10^{-30}\ \text{m}^3]^{1/3}$$

$$= 2.86 \times 10^{-10}\ \text{m}$$

$$= 2.86\ \text{Å}.$$

5. *Calculate the lattice constant for CSCI (bcc lattice). Given: density = 8000 kg/m^3, molecular weight t = 168.4 and N=6.02 × 10^{24} per kg-molecule.*

[**Ans.** 4.11 Å]

6. *The molecular weight of NaCl is 58.5 and its density is 2.17 gm/cm^3. The distance between the adjacent atoms in NaCl crystal is calculated. The Avogadro number is N = 6.02 × 10^{22} per gm-molecule.*

Solution: In the rock-salt (NaCl) crystal, the Na and Cl atoms (strictly ions) occupy alternately the corners of an elementary cube. Let *d* cm be the distance between adjacent atoms.

Then the number of atoms in 1 cm length of an edge of the cube is $1/d$. The total number of atoms in 1 cm^8 (unit volume) is $(1/d)^3$.

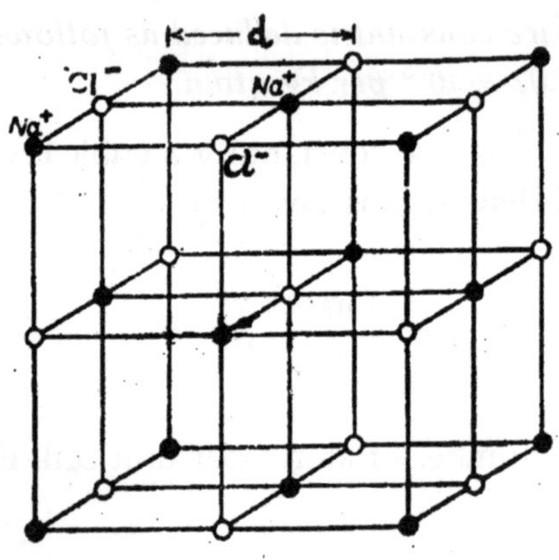

Now, 1 gm-molecule of NaCl contains N molecules and has a mass of M gm, where N is Avogadro's number and M is the molecular weight of NaCl. Thus N molecules of NaCl have a mass M gm and occupy a volume given by

$$V = \frac{M}{\rho}$$

where ρ is the density of NaCl. Thus the number of molecules per unit volume is N/V. Since there are two atoms in each molecule, the total number of atoms per unit volume is $2N/V$. Hence

$$\left(\frac{1}{d}\right)^3 = \frac{2N}{V} = \frac{2N\rho}{M}$$

or $$d = \left(\frac{M}{2N\rho}\right)^{\frac{1}{3}} \text{ cm.}$$

Substituting the given values:

$$d = \left[\frac{58.5 \text{ gm} / (\text{gm - molecule})}{2 \times \left(6.02 \times 10^{23} / \text{gm - molecle}\right) \times 2.17 \text{ gm} / \text{cm}^3}\right]^{\frac{1}{3}}$$

$$= (22.4 \times 10^{-24} \text{ cm}^3)^{1/3}$$

$$= 2.82 \times 10^{-8} \text{ cm} = 2.82 \text{ Å}.$$

7. ***In a crystal, a lattice plane cuts intercepts of a, 2b and 3c along the three axes where a, b, c are primitive vectors of the unit cell. The Miller indices of the given plane is determined as follows.***

Solution: From the law of rational indices, we have

$$a : 2b : 3c = \frac{a}{h} : \frac{b}{k} : \frac{c}{l},$$

where h, k, l are the Miller indices.

Thus

$$\frac{1}{h}:\frac{1}{k}:\frac{1}{l} = 1:2:3$$

$$h:k:l = 1:\frac{1}{2}:\frac{1}{3}.$$

Converting to smallest whole numbers having the same ratios, we get

$$h:k:l = \frac{6}{6}:\frac{3}{6}:\frac{2}{6} = 6:3:2$$

Thus $h = 6$, $k = 3$ and $l = 2$, Hence the Miller indices of the plane are 6, 3 and 2; or the plane is (632).

8. ***The Miller indices of a plane which cuts off intercepts in the ratio 1a : 3b : – 2c along the three axes where a b c are primitives is deduced as follows.***

Solution: From the law of rational indices, we have

$$1a : 3b : -2c = \frac{a}{h}:\frac{b}{k}:\frac{c}{l}$$

where h, k, l are the Miller indices. Then

$$\frac{1}{h}:\frac{1}{k}:\frac{1}{l} = 1:3:-2$$

or
$$h:k:l = 1:\frac{1}{3}:-\frac{1}{2}$$

$$= 6:2:-3.$$

Thus $h = 6$, $k = 2$, $l = -3$. Hence the plane is (623).

9. ***The Miller indices of a set of parallel planes which make intercepts in the ratio 3a : 4b on the X and Y axes and are parallel to the Z axis. a, b, c are the primitive vectors of the lattice is found as follows.***

Solution: The parallel planes are parallel to the Z-axis. This means that their intercepts on the Z-axis are infinite.

From the law of rational indices, we have

$$3a : 4b : \infty\, c = \frac{a}{h} : \frac{b}{k} : \frac{c}{l}$$

or
$$\frac{1}{h} : \frac{1}{k} : \frac{1}{l} = 3 : 4 : \infty$$

$$h : k : l = \frac{1}{3} : \frac{1}{4} : \frac{1}{\infty} = 4 : 3 : 0.$$

The Miller indices are 4, 3, 0.

10. ***In a crystal whose primitives are 1.2 Å, 1.8 Å and 2.0 Å, a plane (231) cuts an intercept 1.2 Å on the X-axis. The corresponding intercepts on the Y- and Z-axis are found as follows.***

Solution: Let p, q and r be the intercepts on the X–, Y–, and Z-axis respectively. Then, from the law of rational indices, we write

$$p : q : r = \frac{a}{h} : \frac{b}{k} : \frac{c}{l}$$

where a, b, c are the primitives and h, k, l are the Miller indices. Here a = 1.2 Å, b =1.8 Å and c = 2.0 Å, and h =2, k = 3 and l = 1. Thus

$$p : q : r = \frac{1.2}{2} : \frac{1.8}{3} : \frac{2.0}{1} = 0.6 : 0.6 : 2.0$$

But p = 1.2 Å.

$$\therefore \quad 1.2 : q = 0.6 : 0.6$$

$$\therefore \quad q = 1.2 \text{ Å}$$

$$\text{Similarly } 1.2 : r = 0.6 : 2.0$$

$$\therefore \quad r = \frac{1.2 \times 2.0}{0.6} = 4.0 \text{ Å}$$

Thus the intercepts on Y and Z axes are 1.2 Å and 4.0 Å respectively.

11. ***The ratio of intercepts on the three axes by (132) planes in a simple cubic lattice is found as follows.***

Solution: Let p, q, r be the intercepts on the X–, Y– and Z–axis respectively. Then,

$$p : q : r = \frac{a}{h} : \frac{b}{k} : \frac{c}{l}$$

where a, b, c are the primitives and h, k, l are the Miller indices. Here $h = 1$, $k = -3$ and $l = 2$; also $a = b = c$ for the "cubic" lattice.

$$\therefore\ p : q : r = \frac{a}{1} : \frac{a}{-3} : \frac{a}{2} = 6 : -2 : 3$$

12. ***The lattice constant for a cubic lattice is a. The spacings between (011), (101) and (112; planes are deduced as follows.***

Solution: For a 'cubic' lattice, the interplanar spacing d is given by

$$d = \frac{a}{\sqrt{(h^2 + k^2 + l^2)}},$$

where h, k, l are the Miller indices.

For (011) plane, $h = 0$, $k = 1$, $l = 1$.

$$\therefore \quad d_{011} = \frac{a}{\sqrt{[(0)^2 + (1)^2 + (1)^2]}} = \frac{a}{\sqrt{2}}$$

Similarly, for (101) plane,

$$d_{101} = \frac{a}{\sqrt{[(1)^2 + (0)^2 + (1)^2]}} = \frac{a}{\sqrt{2}}$$

and for (112) plane,

$$d_{112} = \frac{a}{\sqrt{[(1)^2 + (1)^2 + (2)^2]}} = \frac{a}{\sqrt{6}}$$

13. *The interplanar spacing for a (321) plane in a simple cubic lattice whose lattice constant is 4.2 x 10^{-8} cm is calculated as follows.*

Solution: In a simple cubic lattice the interplanar spacing d is given by

$$d = \frac{a}{\left(h^2 + k^2 + l^2\right)^{1/2}}$$

where h, k, l are the Miller indices. For a (321) plane, we have $h = 3$, $k = 2$ and $l = 1$. Also, here $a = 4.2 \times 10^{-8}$ cm.

$$\therefore \qquad d = \frac{4.2 \times 10^{-8}\,\text{cm}}{\left(3^2 + 2^2 + 1^2\right)^{1/2}}$$

$$= \frac{4.2 \times 10^{-8}\,\text{cm}}{\sqrt{14}} = \frac{4.2 \times 10^{-8}\,\text{cm}}{3.74}$$

$$= 1.12 \times 10^{-8}\text{ cm} = 1.12\text{ Å}.$$

14. *In a tetragonal lattice a = b = 2.5 Å, c = 1.8 Å. The lattice spacing between (111) planes is deduced as follows.*

Solution: The general expression for the interplanar spacing d is

$$d_{hkl} = \frac{1}{\sqrt{\left(\frac{h^2}{a^2} + \frac{k^2}{b^2} + \frac{l^2}{c^2}\right)}}$$

where h, k, l are the Miller indices and a, b, c are the primitives.

Here $h = 1$, $k = 1$, $l = 1$, $a = b = 2.5$ Å, $c = 1.8$ Å.

$$\therefore \qquad d_{111} = \left[\frac{1}{\left(2.5\text{ Å}\right)^2} + \frac{1}{\left(2.5\text{ Å}\right)^2} + \frac{1}{\left(1.8\text{ Å}\right)^2}\right]^{-\frac{1}{2}}$$

$$= \left[\frac{1}{6.25}+\frac{1}{6.25}+\frac{1}{3.24}\right]^{-\frac{1}{2}} \text{Å}$$

$$= [0.16 + 0.16 + 0.31]^{-1/2} \text{ Å}$$

$$= (0.63)^{-1/2} = \left(\frac{100}{63}\right)^{1/2} = 1.26 \text{ Å}$$

15. ***For a simple cubic crystal find (i) the ratio of intercepts on the three axes by (123) plane, (ii) the ratio of the spacings of (110) and (111) planes and (iii) the ratio of the nearest neighbour distance to the next nearest neighbour distance.***

Solution: (*i*) Let p, q, r be the intercepts on the X-, Y- and Z- axis respectively. Then, from the law of rational indices, we have

$$p : q : r = \frac{a}{h}:\frac{b}{k}:\frac{c}{l},$$

where a, b, c are primitives and h, k, l are Miller indices.

Here $h = 1, k = 2, l = 3$; also $a = b = c$ for a "cubic" lattice.

$$\therefore \quad p : q : r = \frac{a}{1}:\frac{a}{2}:\frac{a}{3} = 6 : 3 : 2$$

(ii) The spacing between (hkl) planes in a cubic lattice is given by

$$d_{hkl} = \frac{a}{\left(h^2+k^2+l^2\right)^{1/2}}.$$

Thus $$d_{110} = \frac{a}{\left(1^2+1^2+0^2\right)^{1/2}} = \frac{a}{\sqrt{2}}$$

and $$d_{111} = \frac{a}{\left(1^2+1^2+1^2\right)^{1/2}} = \frac{a}{\sqrt{3}}$$

$$\frac{d_{110}}{d_{111}} = \frac{a/\sqrt{2}}{a/\sqrt{3}} = \sqrt{3}:\sqrt{2}.$$

(iii) In a "simple" cubic lattice the distance between nearest neighbours is a, and that between next nearest neighbours is $\sqrt{2}\,a$ (see Fig.). Their ratio

$$= a/\sqrt{2}a = 1/\sqrt{2}.$$

16. *The densities of lattice points in (111) and (110) planes in a simple cubic lattice is compared.*

Solution: For a simple cubic (primitive) lattice, the (surface) density of lattice points is given by

$$\rho = \frac{d}{abc} = \frac{d}{a^3}, \qquad [\because a = b = c]$$

where d is interplanar spacing. Thus

$$\frac{\rho_{111}}{\rho_{110}} = \frac{d_{111}}{d_{110}}$$

For a simple cubic lattice, $d_{111} = a/\sqrt{3}$ and $d_{110} = a/\sqrt{2}$.

$$\therefore \qquad \frac{\rho_{111}}{\rho_{110}} = \frac{a/\sqrt{3}}{a/\sqrt{2}} = \frac{\sqrt{3}}{\sqrt{2}}$$

or $\qquad P_{111} : P_{110} = \sqrt{2} : \sqrt{3}$

17. *The density of (100) plane in a simple cubic (P) lattice; given a = 2.5 Å.*

Solution: For a cubic P lattice, we have

$$\rho = \frac{d}{a^3}.$$

Now $d_{100} = a$.

$$\therefore \qquad \rho = \frac{a}{a^3} = \frac{1}{a^2}$$

$$= \frac{1}{\left(2.5 \times 10^{-10}\,\text{m}\right)^2}$$

$$= 1.6 \times 10^{19} \text{ lattice point/m}^2.$$

8

Quantum Hypothesis

Photon Theory

Completely Black Body: A perfectly black body is one which absorbs all the thermal radiation incident upon it and, because it does not reflect light, appears black. No existing body is perfectly black. An object coated with a diffuse layer of black pigment is nearly a black body.

Black Body in Practice: A cavity in a body, connected to the outside by small hole, is a practical black body (Fig.). Radiation outside the cavity enters it through the hole. This radiation strikes the inner wall (1) v, here it is partly absorbed and partly reflected. The reflected part strikes another part of the inner wall (2) and the process is repeated, until the incident radiation is totally absorbed by the wall.

Since the area of the hole is negligible compared to the total area of the cavity wall, the part of the incoming radiation that can get out through the hole can be neglected. Thus all the radiation incident on the hole from the outside, is absorbed by it. Hence the hole behaves as a black body. At low temperature the hole appears black.

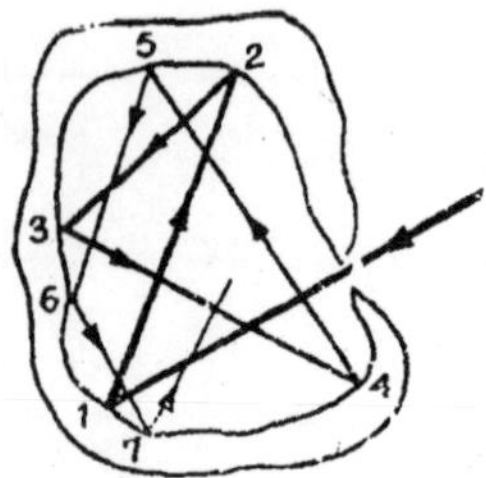

Black-Body Radiation and its Characteristics: The radiation emitted by a *heated* black body is called 'black-body radiation.' If the walls of the cavity shown in Fig. are heated *uniformly,* the hole becomes self-luminous. The inner walls emit thermal radiation into the cavity which undergoes many successive reflections and is eventually re-absorbed. In the state of thermal equilibrium the emission rate is equal to the absorption rate, and a constant amount is in transit inside the cavity. Some very small part of this radiation emerges through the hole. Since the hole behaves as a black-body, the radiation emitted by it is the black-body radiation. It is also known as cavity radiation'. It has following important characteristics:

(*i*) *The cavity radiation is more intense than the radiation from a non-black body at the same temperature.* This is why the hole appears brighter than the outer wall (non-black body) of the cavity.

(*ii*) *At a given temperature, the cavity radiancy is independent of the material, shape and size of the cavity.* If this were not so, then on joining together two cavities *A* and *B* made of different materials but heated to the same temperature, a net amount of radiant energy would have flown from *A* to *B* (say). This would heat up *B* and cool down *A*. Then *B* would be used as the source and *A* as the sink of a heat-engine and work could be obtained until *A* and *B* were again at the same temperature. This process could be repeated as many times as desired, thus obtaining a continuous supply of work. This is clearly a violation of the second law of thermodynamics. Hence we conclude that the radiancy of two cavities at the same temperature but of different materials cannot be different from each other. (On the other hand, the radiancy of the outer surface does depend upon the material).

(*iii*) *The cavity radiancy E is directly proportional to the fourth power of the Kelvin temperature T of the cavity* (Stefan's law). That is

$$E = \sigma T^4$$

where σ is the universal Stefan-Boltamann constant. (On the other hand, the radiancy of the outer surface is $e\sigma T^4$, where *e* is the 'emissivity' which depends upon the material and upon the temperature).

(*iv*) *The spectral distribution of energy in the black body radiation at a given temperature is independent of the material, shape and size of the body.*

What is Cavity Hole ?

A black-body is one that completely absorbs all the radiation, whatever be the wavelength, falling on it and reflects none. Now, any radiation from outside falling on the fine hole made in the walls of a cavity radiator is completely absorbed by successive reflections at the inner walls and none is reflected back to the eye. Hence, the hole behaves as a black-body.

Interior of the Cavity Radiator: The hole of a cavity radiator behaves as a black-body. On looking into it, it reflects no light into the observer's eye and appears as black. Hence the details of the interior of the cavity are not seen.

Coal Pockets Brighter: The radiancy of the hole of a cavity radiator, given by σT^4, is larger than the radiancy of the outer surface which is given by $e\sigma T^4$, where e is the surface emissivity and is less than 1. Therefore, the radiation coming out from the 'pockets' formed by the glowing coal is more intense than that coming from the surface of the coals, although both are at the same temperature. Hence the pockets appear brighter than the coal themselves.

Law of Stefan

The Stefan's fourth power law ($E = \sigma T^4$) is exact for true black bodies only. Now, only certain incandescent bodies, like platinum black, are near a true black body. Others are far away. Hence all incandescent bodies do not obey this law.

Stefan's Law for Defining Temperature: A body heated to a temperature say 50° C, cannot be treated as a black body and so the Stefan's law will not be applicable to it. Hence we cannot use this law for defining this temperature.

Spectral Distribution of Energy in Black-Body Radiation: The distribution of energy among the various wavelengths in black-body radiation was investigated by Lummer and Pringsheim in 1899. They used an electrically-heated chamber with a small aperture as the black body and measured its temperature by a thermocouple. The whole arrangement is shown in Fig.

The radiation from the black body O is made to fall on a slit *S* by means of a concave mirror .4. The slit *S* is in the focal plane of a second concave mirror *B* which reflects the radiation as a parallel beam which falls on a fluor spar prism *P.* The emerging beam falls on a third concave mirror *C* which directs it on to a Lummer-Kurlbaum linear bolometer *T.* On rotating the mirror *C* about a vertical axis, the radiations of different wavelengths fall on the bolometer one after the other.

The deflection of the galvanometer gives the corresponding spectral radiancy $E\lambda$, which is defined such that the quantity $E\lambda \, d\lambda$ is the energy, for wavelengths lying between A and $\lambda + d\lambda$, emitted per second per unit surface area of the black body.

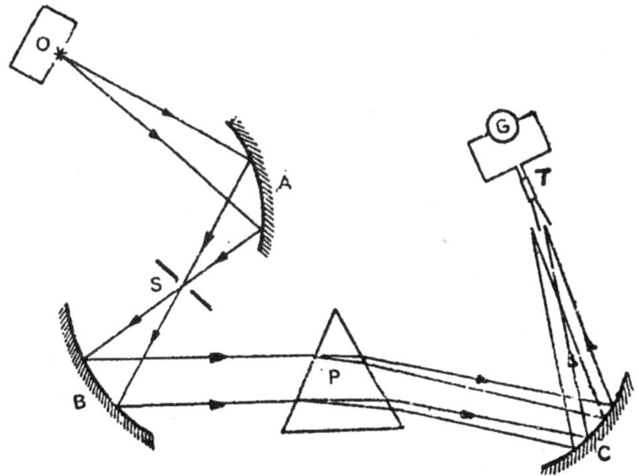

Fig. shows observed spectral radiancy $E\lambda$ plotted against the wavelength A of radiation at three different black-body temperatures. These curves show three important features:

(*i*) The total energy emitted per second per unit area (*i.e.* radiancy E or area under the curve) increases rapidly with increasing temperature T. THE increase is found in accordance with the Stefan's law:

$$E = \int_0^\infty E_\lambda \, d\lambda = \sigma T^4.$$

(*ii*) At a particular temperature, the spectral radiancy E_λ is a maximum at a particular wavelength λ_m (say). Most of the energy is emitted at wavelengths not very different from λ_m

(*iii*) The wavelength λ_m for maximum spectral radiancy decreases in direct proportion to the increase in temperature. This is called 'Wien's displacement law' According to this law,

$$\lambda_m \times T = \text{constant}.$$

This means that as the temperature is raised, the cavity emits more and more the radiation of the shorter wavelengths.

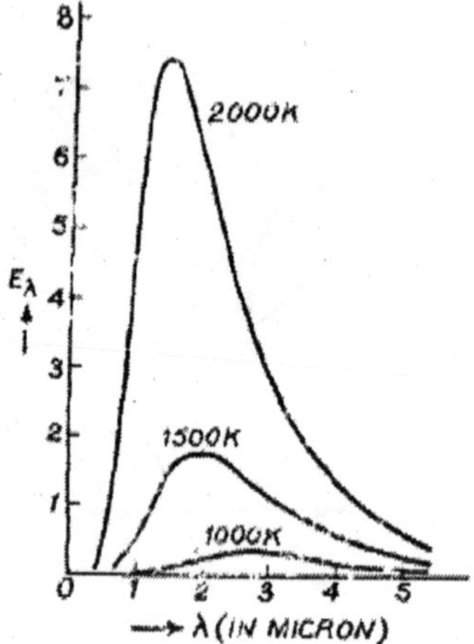

Success and Limitations of Classical Theory

Wien's and Rayleigh-Joans Formulae: First of all, Wien in 1893, worked on the distribution of energy among the wavelengths. He showed from pure thermodynamics reasoning that the energy-density $u\lambda$ in the wavelength-interval λ to $\lambda + d\lambda$ emitted by a black body at temperature T is of the form

$$d\lambda \; d\lambda = \frac{A}{\lambda^5} \; \mathrm{f} \; (\lambda T) \; d\lambda,$$

where A is a constant and $f(\lambda T)$ is an undetermined function. This is in agreement with the experimental result that the product λT at the peak of the $F\lambda$-λ curve is same at all temperatures.

In order to find the form of $f(\lambda T)$, Wien assumed that the black-body radiation inside a cavity may be supposed to be emitted by resonators of molecular dimensions having. Maxwellian velocity distribution, and the frequency of emitted radiation is proportional to the kinetic energy of the corresponding resonator.

On this basis, Wein established the following distribution formula:

$$u\lambda \, d\lambda = \frac{A}{\lambda^5} e^{-B/\lambda T} d\lambda,$$

where A and B are constants.

Wein's formula was found to agree with experiment at short wave-lengths but did not fit well at long wavelengths (Fig.). According to the formula $u\lambda = 0$ for $\lambda = 0$ and also for $\lambda = \infty$ as it should be, but it keeps $u\lambda$ finite for $T = \infty$ which is unlikely.

Rayleigh and Jeans considered the black-body radiator (cavity) full of electromagnetic waves of all wavelengths between 0 and ∞ which, due to reflection at the walls, form standing waves. They calculated the number of possible waves having wavelengths between A and $\lambda + d\lambda$, and using law of equipartition of energy, established the following distribution formula:

$$u\lambda \, d\lambda = \frac{8 \pi k T}{\lambda^4} d\lambda,$$

where k is the Boltzmann's constant,

Rayleigh-Jeans formula was found to agree with experiment at long wavelengths only (Fig.). It is, however, open to a serious objection. As we move towards shorter wavelengths (*i.e.* towards ultraviolet), the predicted energy would increase *without limit,* thus diverging enormously from experiment. This completely erroneous prediction is known as the 'ultraviolet catastrophe;'.

Furthermore, at any temperature T the total energy of radiation, $\int_0^\infty \frac{8 \pi k T}{\lambda^4} d\lambda$, is predicted to be infinite which is against the Stefan's law.

The shortcomings of classical theory were overcome by Planck's quantum hypothesis is given below.

Explanation of Planck's hypothesis of quantum theory of radiation. Let's deduce an expression for the average energy of a Planck's oscillator and obtain Planck's radiation formula, the Rayleigh-Jean's law and Wien's law are special cases of Planck's law is proved here.

Hypothesis of Planck

Planck, in 1900, intro-entirely new ideas to explain the distribution of energy among the various wavelengths of the cavity radiation. He assumed that the *atoms of the walls of the cavity radiator behave as oscillators, each with a characteristic frequency of oscillation.* These oscillators emit electromagnetic radiant energy into the cavity and also absorb the same from it, and maintain an equilibrium state. Planck made two rather revolutionary assumptions regarding these atomic oscillators:

(*i*) *An oscillator can have only discrete energies given by*

$$\in = nhv,$$

where v is the frequency of the oscillator, *h* is a constant known as 'Planck's constant', *n* is an integer known as 'quantum number'. This means that the oscillator can have only the energies *hv*, 2*hv*, 3*hv*,...and not any energy in between. In other words, the energy of the oscillator is quantised.

(*ii*) *The oscillators do not emit or absorb energy continuously but only in 'jumps'*. That is, an oscillator emits or absorbs packets of energy, each packet carrying an amount of energy *hv*.

$$\Delta \in = (\Delta n)\, hv;$$

$$\Delta n = 1.\ 2,...$$

Average Energy of Planck's Oscillator: Let us now calculate the average energy of Planck's oscillator of frequency v. The relative probability that an oscillator has the energy *hv* at temperature *T* is given by the Boltzmann, factor $e^{-hv/kT}$. Now, let $N_{0,}$ N_1 N_2...N_r..be the number of oscillators having energies

0, *hv*, 2*hv*...*rhv*...respectively. Then we have $N_r = N_0\, e^{-hv/kT}$. The total number of oscillators is

$$N = N_0 + N_1 + N_2 + \dots \dots$$
$$= N_0\,(1 + e^{-hv/kT} + e^{-2hv\,/\,kT} + \dots \dots)$$
$$= \frac{N_0}{1 - e^{-hv/kT}}. \qquad \dots(i)$$

The *total* energy of the oscillators is given by

$$\in\, = (N_0 \times 0) + (N_1 \times hv) + (N_2 \times 2hv) + \dots\dots$$
$$= (N_0 \times 0) + (N_0\, e^{-hv\,/\,kT} \times hv) + (N_0\, e^{-2hv\,/\,kT} \times 2hv) + \dots \dots$$
$$= N_0 e^{-hv\,/\,kT}\, hv(1 + 2e^{-hv\,/\,kT} + 3e^{-2hv\,/\,kT} + \dots \dots)$$
$$= N_0 e^{-hv\,/\,kT} \frac{hv}{\left(1 - e^{-hv/kT}\right)^{2*}} \qquad \dots (ii)$$

Dividing eq. (ii) by eq. (i), we obtain the *average* energy of an oscillator as given by

$$\bar{\in} = \frac{\in}{N} = \frac{e^{-hv/kt}}{1 - e^{-hv/kT}} = \frac{hv}{e^{hv/kT-1}}. \qquad \dots(iii)$$

This is the expression for the average energy of a Planck's oscillator.

Planck's Radiation Formula: The energy-density of radiation u_v in the frequency range v to v + *dv* is related to the average energy of an oscillator emitting *v*-radiation by

$$u_y\, d_v = \frac{8\,\pi\, v^2}{c^3}\, dv \times \bar{\in}.$$

Substituting the value of $\bar{\in}$ from eq. (iii), we have

$$u_v\, dv = \frac{8\,\pi\, v^2}{c^3} \frac{hv}{e^{hv/kT} - 1}\, dv$$

or

$$u_v\, dv = \frac{8\,\pi\, h}{c^3} \frac{v^3\, dv}{e^{hv/kT} - 1}.$$

This is Planck's radiation formula in terms of frequency v to express it in terms of wavelength we observe that, since

$v = \frac{c}{\lambda}$,

$$dv = -\frac{c}{\lambda^2}\, d\lambda$$

and since an increase in frequency corresponds to a decrease in wavelength,

$$u_\lambda \; d\lambda = -\, u_v \; dv$$

Therefore $$u_\lambda \; d\lambda = \frac{8\pi h}{c^3} \frac{\left(\frac{c}{\lambda}\right)^3 \left(\frac{c}{\lambda^2}\, d\lambda\right)}{e^{hc/\lambda kT} - 1}$$

or $$u_\lambda \; d\lambda = \frac{8\pi hc}{\lambda^5} \frac{1}{e^{he/\lambda kT} - 1} \qquad \text{... (iv)}$$

This is Planck's formula in terms of wavelength.

Explanation of Energy Distribution by Planck's Formula: Wien's Law and Rayleigh-Jeans Law are Special Cases: The Planck's formula is found to be in complete agreement with experiment for the entire wavelength range at all temperatures. This can be seen in the following way:

(1) When λ is very small, then $e^{hc/\lambda kT} >> 1$, so that Planck's formula given by eq. (IV) can be written as

$$u_\lambda \; d\lambda = \frac{8\pi hc}{c^5}\, e^{-\,hc/\lambda kT}\, d\lambda.$$

Putting $8\,\pi hc = A$ and $\frac{hc}{k} = B$, we have

$$u_\lambda \; d\lambda = \frac{A}{\lambda^5}\, e^{-B/\lambda T}\, d\lambda.$$

This is Wien's law which agrees with experiment at short wavelengths.

(2) When λ is very large, then $e^{hc/kT} \approx 1 + \dfrac{hc}{\lambda kT}$ so that (iv) can be written as

$$u_\lambda \; d\lambda = \frac{8\,\pi\, hc}{\lambda^5\left(1+\dfrac{hc}{\lambda kT}-1\right)} d\lambda$$

$$= \frac{8\,\pi\, k\, T}{\lambda^4}\; d\lambda$$

This is Rayleigh-Jeans law which agrees with experiment at long wavelengths.

Further, both Wien's displacement law and Stefan's law can be derived from the Planck's formula.

Wien's Displacement Law from Planck's Formula: The Planck's radiation formula is

$$u_\lambda \; d\lambda = \frac{8\,\pi\, hc}{\lambda^5}\frac{d\lambda}{e^{hc/\lambda kT}-1}$$

or $$u_\lambda = 8\pi hc\; (\lambda^{-5})\; (e^{hc/\lambda kT} - 1)^{-1}.$$

To find the wavelength at which the spectral radiancy is maximum, we put

$$\frac{du_\lambda}{d\lambda} = 0$$

that is

$$8\pi hc\left[-5(\lambda^{-6})\,(e^{hc/\lambda kT}-1)^{-1} + \lambda^{-5}\,(-1)\,(e^{hc/\lambda kT}-1)^{-2}\, e^{hc/\lambda kT}\,\frac{-hc}{\lambda^2 kT}\right] = 0$$

or $$\frac{5}{\lambda} = (e^{hc/\lambda kT} - 1)^{-1}\; e^{hc/\lambda kT}\,\frac{hc}{\lambda^2\, kT}$$

or $$5 = \frac{hc}{\lambda kT}\frac{e^{hc/\lambda kT}}{e^{hc/\lambda kT}-1}$$

Putting $$\frac{hc}{\lambda kT} = x, \text{ we get}$$

$$5 = \frac{xe^x}{e^x - 1} = \frac{x}{1 - e^{-x}}$$

or $$\frac{x}{5} + e^{-x} = 1.$$

This equation has a single root given by $x = 4.965$, and therefore x must be a constant. That is

$$\frac{hc}{\lambda kT} = 4.965$$

Therefore, the wavelength λ_m at which the spectral radiancy per unit range of wavelength has its maximum value is given by

$$\lambda_m T = \frac{hc}{4.965\, k} = b \text{ (say)}$$

This is Wien's displacement law.

Stefan's Law from Planck's Formula: Let us write the Planck's radiation formula in terms of frequency;

$$u_v \, dv = \frac{8\pi h}{c^3} \frac{v^3 \, dv}{c^{hv/kT} - 1}$$

The spectral radiancy E_v is related to the energy-density u, by

$$E_v = \frac{c}{4} u_v$$

Thus $$E_v \, dv = \frac{2\pi h}{c^2} \frac{v^2 \, dv}{e^{hv/kT} - 1}$$

The total radiant energy over all frequencies is

$$E = \int_0^\infty E_v \, dv = \frac{2\pi h}{c^2} \int_0^\infty \frac{V^3 \, dv}{e^{hv/kT} - 1}$$

Let us put $\frac{hv}{kT} = x$, so that $v = \frac{Kt}{h} x$ and $dv = \frac{Kt}{h} dx$. Then

$$E = \frac{2\pi k^4 T^4}{h^3 c^2} \int_0^\infty \frac{x^3 \, dx}{e^{x-1}}.$$

The value of the integral $\int_0^\infty \frac{x^3 \, dx}{e^{x-1}}$ is $\frac{\pi^4}{15}$. Thus

$$E = \frac{2\pi^5 k^4}{15 h^3 c^2} T^2$$

Let us put $\frac{2\pi^5 k^4}{15 h^3 c^2} = \sigma$ (a universal Constant). Then

$$E = \sigma T^4.$$

This is Stefan's law.

Importance of Photon

The different forms of energy such as radio waves, heat rays, ordinary light, X-rays, y-rays, etc., which are emitted by atoms under different situations are all electro-magnetic radiations varying in frequency (or wavelength). We call the electro-magnetic *waves* because under suitable circumstances they exhibit refraction, interference and diffraction.

The wave-like character of radiation, however, failed to explain the observed energy distribution in the continuous spectrum of radiation emitted by hot bodies. To meet this serious problem, Planck, in 1901, presented his quantum theory of radiation. According to this theory, radiation is emitted (or absorbed) *discontinuously* in *indivisible* packets of energy. These packets were named as 'photons' or 'Quantas. Each photon of radiation of a given frequency v has the same energy which is hv, where h is now known as Planck's constant. Planck did not suggest anything new regarding the propagation of radiation in space.

Einstein extended Planck's quantum hypothesis by assuming that *radiation not only is emitted (or absorbed) as indivisible photons, but also continues to propagate through space as photons.* On this hypothesis he in 1905 explained photoelectric effect and in 1907 tackled the problem of specific heat of solids. In 1913, Bohr used the quantum theory to explain the hydrogen spectrum and in 1922, Compton applied it to the scattering of X-rays.

Properties of Photons

(i) Photons are *indivisible* packets of electromagnetic energy.

(ii) They retain their identity until completely absorbed by some atom (as happens in photoelectric effect and pair production).

(iii) The size (energy content) of a photon (hv) is proportional to the frequency of radiation so that photons of different radiations are of different sizes. For example, blue photons are larger than red photons; X-ray photons are considerably larger than visible light photons.

(iv) The intensity of radiation (I) is equal to the number of photons (N) crossing unit area per second multiplied by the size (hv) of the photon, *i.e.*,

$$I = N\,hv.$$

Thus for a given frequency the intensity depends simply upon the number of photons.

(v) *All* photons travel with the speed of light in vacuum (,) and have a zero rest mass. Hence there is no frame in which a photon is at rest.

(vi) From the special theory of relativity, the total energy E of a particle is related to its rest mass m_0 and momentum p by the relation

$$E = \sqrt{(m_0^2c^4 + p^2c^2)}.$$

For a Photon, $m_0 = 0$; therefore

$$E = pc.$$

But $E = h\upsilon$. Thus the photon has a momentum given by

$$p = \frac{E}{c} = \frac{h\upsilon}{c}$$

The Compton effect is a direct evidence of the existence of momentum for a photon.

(vii) The existence of momentum for a photon implies that it must have an effective mass m also. This mass can be computed by mass-energy relation ($E = mc^2$).

$$m = \frac{E}{c^2} = \frac{h\upsilon}{c^2}.$$

Photons and Corpuscles: The particle-like photon picture of radiation appears to be a return to the Newton's corpuscular theory of light. It, however, is not quite so. Corpuscles were material particles (with finite rest mass) whereas photons are energy packets with zero rest mass. Moreover, the energy of photons is related to the frequency of radiation. The photons, unlike corpuscles, include the wave picture of radiation also.

Complementarity of Waves and Corpuscles: In certain events radiation shows a wave-like character while in others it shows a particle-like character. The same light beam which is diffracted by a grating (showing wave-like character) can cause the emission of photo electrons from a suitable surface (showing particle-like character).

These processes, however, occur independently. Thus the wave and particle aspects of radiation complement each other. Hence the radiation is supposed to have a dual character behaving as a wave in one situation and as a particle in the other situation.

PROBLEMS

1. *Find the wavelength at which the spectral radiancy of a cavity radiator at 6000 K is maximum.*

Solution: By Wien's displacement law, the wavelength λ_m for the maximum radiation is given by

$$\lambda_m T = \frac{hv}{4.965\,k}.$$

Putting $h = 6.62 \times 10^{-34}$ joule-sec, $c = 3.0 \times 10^{8}$ meters /sec and $k = 1.38 \times 10^{-23}$ joule / K, we get

$$\lambda_m T = \frac{\left(6.62 \times 10^{34}\right) \times \left(3.0 \times 10^{8}\right)}{4.965 \times \left(1.38 \times 10^{-23}\right)}$$

$$= 2.898 \times 10^{-3} \text{ Meter -K.}$$

For $T = 6000$ K, we get

$$\lambda_m = \frac{2.898 \times 10^{-3}}{6000}$$

$$= 0.483 \times 10^{-6} \text{ meter-K.}$$

$$= 4830 \text{ Å.} \qquad [1\text{A} = 10^{-10} \text{ meter}]$$

2. *From Planck's law after deducing the value of v corresponding to the peak of the $E_v - v$ curve for black body radiation at 1000 K. ($h = 6.62 \times 10^{-34}$ joule-sec, $k = 138 \times 10^{-23}$ joule/K) The result is as follows.*

Solution: The wavelength λ_m at which the radiant energy is maximum, is given by

$$\lambda_m T = \frac{hc}{4.965\,k}$$

If v be the frequency corresponding to λ_m, then

$$\text{v} = \frac{c}{\lambda_m} = \frac{4.965\,kT}{h}$$

Here T = 1000 K.

$$\therefore \qquad v = \frac{4.965 \times \left(1.38 \times 10^{23} \text{ joule / K}\right)\left(1000 \text{ K}\right)}{6.62 \times 10^{-84} \text{ joule} - \text{sec}}$$

$$= 1.035 \times 10^{14} \text{ sec}^{-1}$$

3. The maximum in the energy distribution spectrum of the sun is at 4735 Å and its temperature is 6050 E. The temperature of a star whose energy distribution shows a maximum at 9506 Å will be as follows.

Solution: By Wein' s displacement law,

$$\lambda_m T = \text{constant.}$$

Applying it for the sun and the star, we have

$$4735 \text{ Å} \times 6050 \text{ K} = 9506 \text{ Å} \times T$$

or
$$T = \frac{4735 \text{ Å} \times 60.50 \text{ K}}{9506 \text{ Å}}$$

$$= 3013 \text{ K.}$$

4. A black body at 400 K radiates energy at the rate of 1.45 × 10³ wstt/m². We can try to obtain a value for Stefan's constant.

Solution: By Stefan's law, we have

$$E = \sigma T^4$$

Substituting the given values of E and T, we have

$$\sigma = \frac{E}{T^4}$$

$$= \frac{1.45 \times 10^2 \text{ watt / m}^2}{\left(400 \text{ K}\right)^4}$$

$$= 5.66 \times 10^{-8} \text{ watt/(m}^2 - \text{K}^4\text{).}$$

5. The surface temperature of the star which is radiating 1.6×10^6 times more energy per unit area per second compared to sun may be evaluated here. Surface temperature of the sun is 6000 K.

Solution: Let T be the surface temperature of the star. Then by Stefan's law, we have

$$1.6 \times 10^5 = \frac{T^4}{(6000\text{ K})^4}$$

$$\text{or} \qquad T^4 = 1.6 \times 10^5 \times (6000\text{ K})^4$$

$$= 16 \times 36 \times 36 \times 10^{16}$$

$$\therefore \qquad T = 120000\text{ K}.$$

6. Let us calculate the energy of a photon of ultraviolet light of wavelength 1800 A and of infrared light of wavelength 36 μ, ($h = 6.63 \times 10^{-27}$ erg sec, $c = 30 \times 10^{10}$ cm/sec and 1 eV = 1.6×10^{-12} erg).

Solution: The energy of a photon of frequency v is

$$E = hv = h\frac{c}{\lambda},$$

where c is the velocity of light and λ the wavelength. Here $\lambda = 1800$ Å $= 1800 \times 10^{-8}$ cm.

$$\therefore \qquad E = \frac{(6.63 \times 10^{-27}\text{ erg} - \text{sec}) \times (3.0 \times 10^{10}\text{ cin sec}^{-1})}{1800 \times 10^{-8}\text{ cm}}$$

$$= 1.1 \times 10^{-11}\text{ erg}$$

$$= \frac{1.1 \times 10^{-11}}{1.6 \times 10^{-12}}$$

$$= 6.9\text{ eV}.$$

For $\lambda = 3.6\ \mu = 3.6 \times 10^{-4}$ cm, we can see that

$$E = 0.345\text{ eV}.$$

7. The wavelength of a 100-MeV photon. (h = 6.63 × 10^{-34} joule-sec, c = 3 × 10^8 meters-sec-^{1}, and 1 eV = 1.6 ×10^{19} joule.) may be calculated

Solution: The energy of the photon is given by

$$E = hv = \frac{hc}{\lambda}$$

$$\therefore \qquad \lambda = \frac{hc}{E}$$

Here E = 100 MeV = 100 × 10^c eV = 10^8 × (1.6 × 10^{-10}) joule.

$$\therefore \qquad \lambda = \frac{\left(6.63 \times 10^{-34} \text{ joule-sec}\right) \times \left(3 \times 10^8 \text{ meters sec}^{-1}\right)}{10^8 \times \left(1.6 \times 10^{19}\right) \text{joule}}$$

$$= 124 \times 10^{-14} \text{ meter} = 1.24 \times 10^{-4} \text{ Å}$$

8. We can calculate in A the wavelength of a photon whose quantum energy is equal to the rest energy of an electron (m_0 = 9.1 × 10^{-31} kg, h = 6.62 × 10^{-34} joule-sec, c = 3 × 10^8 meter/sec).

Solution: The rest energy of electron is w_0c^2. The wavelength of a photon whose energy is $E = w_0c^2$ is given by

$$\lambda = \frac{hc}{E} = \frac{h}{m_0 c}$$

Putting the given values:

$$\lambda = \frac{6.62 \times 10^{-34}}{\left(9.1 \times 10^{-31}\right) \times \left(3 \times 10^8\right)}$$

$$= 0.024 \times 10^{-10} \text{ meter} = 0.024 \text{ Å}.$$

9. A faint star is just visible if 2000 photons/sec enter the eye. What energy per second (in watts) this represents in terms of sodium light is given below (λ = 5890 Å). (h = 6.63 × 10^{-34} joule-sec and c = 3 × 10^8 m/sec).

Solution: The energy of a photon is

$$E = hv = \frac{hc}{\lambda}.$$

Here $\lambda = 5890\ \text{Å} = 5890 \times 10^{-10}$ meter.

$$\therefore \quad E = \frac{\left(6.63 \times 10^{-34}\right) \times \left(3 \times 10^{8}\right)}{5890 \times 10^{-10}}$$

$$= 3.38 \times 10^{-19} \text{ joule.}$$

The number of photons entering the eye is 2000 per sec. Hence the total energy received

$$= 2000 \times (3.38 \times 10^{-19})$$

$$= 676 \times 10^{-10} \text{ joule/sec}$$

$$= 6.76 \times 10^{-16} \text{ watt.}$$

10. A 1000-watt radio transmitter operates at a frequency of 880 kc/sec. We can see as to how many photons per second it emits ($h = 6.63 \times 10^{-34}$ joule-sec).

Solution: The electromagnetic energy radiated by the 1000-watt transmitter is 1000 joule/sec. The energy of a photon of radiation of frequency v (880×10^3/sec) is

$$E = hv = (6.63 \times 10^{-34} \text{ joule-sec}) \times (880 \times 10^{3} \text{ sec}^{-1})$$

$$= 5.83 \times 10^{-28} \text{ joule.}$$

Hence the number of photons emitted per sec

$$= \frac{\text{total energy emitted per sec}}{\text{energy of one photon}}$$

$$= \frac{1000 \text{ joules per sec}}{5.83 \times 10^{-28}} = 1.71 \times 10^{30} \text{ per sec.}$$

11. If a 60-watt light source is emitting light of wavelength 5400 A. The rate of emission of photons from the source would be as follows.

Solution: The energy of a photon of light of wavelength λ = 5400 Å is

$$E = hv = \frac{hc}{\lambda}$$

$$= \frac{(6.63 \times 10^{34} \text{ joule-sec}) \times (3.0 \times 10^{8}\ m - sec^{-1})}{(5400 \times 10^{10} \text{ m})}$$

$$= 3.68 \times 10^{-19} \text{ joule.}$$

The rate of energy emission from a 60-watt lamp is 60 joules/sec. Hence the number of photons, each of energy 3.68×10^{-19} joule, emitted per second is $\frac{60}{3.68 \times 10^{-19}} = 1.63 \times 10^{20}$.

12. Solar radiation falls on the earth at a rate of 1340 watt/m² on a surface normal to the incoming rays. Assuming that sunlight consists exclusively of 5500 Å photons, the volume of photons per meter² per sec are reaching the earth is as follows.

Solution: The energy of solar radiation is 1340 watts/m^2 or 1340 joules/m^2-sec. The energy of a 5500-Å photon is

$$E = \frac{hc}{\lambda}$$

$$= \frac{(6.63 \times 10^{34} \text{ joule-sec}) \times (3.0 \times 10^{8}\ m - sec^{-1})}{(5500 \times 10^{-10} \text{ m})}$$

$$= 3.616 \times 10^{-19} \text{ joule.}$$

The number of these photons reaching the earth is $\frac{1340}{3.616 \times 10^{-19}} = 3.71 \times 10^{21}$ photons/m^2- sec.

13. Radiation of wavelength 5000 A and intensity 2 × 10⁻² watt/cm² falls on a photosensitive surface. Assuming that every photon is absorbed and results in the ejection of

a photoelectron, the way of finding the number of photo electrons are produced per cm² per sec is as follows. (h = 6.63 × 10^{-27} erg-sec, c = 3 ×10^{10} cm/sec).

Solution: The intensity of radiation is 2×10^{-2} watt/cm² or $2 \times 10^{-2} \times 10^{7}$ ergs*/cm²-sec. The energy of a 5000-Å photon is

$$E = \frac{hc}{\lambda}$$

$$= \frac{\left(6.63 \times 10^{-27} \text{ erg-sec}\right)\left(3 \times 10^{10} \text{ cm} - \text{sec}^{-1}\right)}{\left(5500 \times 10^{-8} \text{ cm}\right)}$$

$$= 3.978 \times 10^{-12} \text{ erg.}$$

∴ no of photons

$$= \frac{\text{intensity of radiation}}{\text{photon energy}}$$

$$= \frac{2 \times 10^{-2} \times 10^{7}}{3.978 \times 10^{-12}}$$

$$= 5.03 \times 10^{16} \text{ / cm}^2 \text{ - sec.}$$

Since every absorhed photon is ejecting a photoelectron, the number of photo electrons produced per cm² per second is 5.03×10^{16}.

14. The frequency of an x-ray photon whose momentum is 1.1 × 10^{-18} gm-cm/sec is explained below (h = 6.63 × 10^{-27} erg-sec and c = 3 × 10^{10} cm-sec^{-1}.)

Solution: The energy of a photon of radiation of frequency v is hv, and its momentum p is hv/c. Here

$$p = \frac{hv}{c}$$

$$= 1.1 \times 10^{-18} \text{ gm-cm/sec.}$$

$$\therefore \qquad v = \frac{c}{h} \times p$$

$$= \frac{3 \times 10^{10}}{6.63 \times 10^{-27}} \times (1.1 \times 10^{-28})$$

$$= 5 \times 10^{18} \text{ cycles/sec.}$$

15. An x-ray photon has a wavelength 0.2 Å. Its energy (in eV) and momentum may be computed.

Solution: The energy E of a photon of frequency v is given by

$$E = hv = \frac{hc}{\lambda},$$

where λ is the wavelength.

Here A = 0.2 Å = 0.2 × 10^{-8} cm.

$$\therefore \qquad E = \frac{\left(6.63 \times 10^{-27} \text{ erg} - \text{sec}\right) \times \left(3 \times 10^{10} \text{ cm} - \text{sec}^{-1}\right)}{\left(0.2 \times 10^{8} \text{ cm}\right)}$$

$$= 9.945 \times 10^{-8} \text{ erg}$$

$$= \frac{9.945 \times 10^{-8}}{1.6 \times 10^{-12}} \qquad [\because 1 \text{ eV} = 1.6 \times 10^{-12} \text{ erg}]$$

$$= 6.2 \times 10^{4} \text{ eV.}$$

The momentum p of a photon of frequency v is given by

$$p = \frac{hv}{c} = \frac{E}{c}$$

$$= \frac{9.94 \times 10^{-8} \text{ erg}}{3 \times 10^{10} \text{ cm - sec}^{-1}}$$

$$= 33.15 \times 10^{-18} \text{ gm-cm/sec}$$

17. The effective mass of a photon of wavelength 10 Å ($h = 6.62 \times 10^{-34}$ joule-sec and $c = 3 \times 10^8$ m/sec) may be computed.

Solution: A photon is a particle of zero rest mass moving at the speed of light and having a finite energy given by

$$E .= mc^2.$$

Its effective' mass *in* is therefore

$$m = E/c^2.$$

For a photon of wavelength.;

We have $E = hv = hc/\lambda$.

$$\therefore \qquad m = \frac{h}{c\lambda}$$

$$= \frac{6.62 \times 10^{-34} \text{ joule-sec}}{\left(3 \times 10^8 \text{ m/sec}\right) \times \left(1.0 \times 10^{-10} \text{ m}\right)}$$

$$= 2.21 \times 10^{-22} \text{ kg.}$$

9

Integrated Circuits

Microelectronic Circuits

Benefits of Integrated Cicuits: We are familiar with the 'discrete circuits' consisting of separately manufactured active and passive circuit components, externally inter-connected by wires. The circuits are called 'discrete', because each component of the circuit is discrete from the others. These circuits occupy large space, and have a number of joints which make them somewhat unreliable. To meet the military requirement of miniature (mini) electronic equipment, a new branch of electronics called 'microelectronics' originated in the late 1950s. It deals with microelectronic circuits called as "integrated circuits", abbreviated as ICs. In an IC all components like resistors, capacitors, diodes, transistors, etc. are fabricated on a monolithic (single) semiconductor chip. Thus an integrated circuit is a packaged electronic circuit which has the advantages of high reliability, small physical size, low cost, and low power consumption.

The individual components of an IC can, however, neither be removed nor be replaced because each one of them is an integral part of the same semiconductor chip. A typical size of a semiconductor chip is 50 mils × 50 mils × 5 mils (1 mil = 0.001 inch).

Integrated Circuits: Kinds

Monolithic ICs: A monolithic IC is one in which all circuit components and their interconnections are formed on a single semiconductor chip. Monolithic circuit components include resistors, capacitors, transistors (BJTs and FETs).

Film ICs: A film IC is one in which the circuit components are formed on a substrate. The components include resistors, capacitors and thin-film transistors.

Hybrid ICs: A hybrid IC is a combination of two or more ICs or one IC and several discrete components.

Multi-chip ICs: In a multi-chip IC, the circuit components are fabricated on separate chips which are attached to a substrate and interconnected as if they were discrete components. These circuits have limited use.

Integrated Circuits (ICs): A (monolithic) integrated circuit consists of a single silicon chip in which both active and passive components have been diffused, and interconnected by aluminium metallization.

Fabrication of a Monolithic IC: With the exception of inductance, all the electronic circuit components (resistors, capacitors, diodes, junction and field-effect transistors) can be formed in semiconductor form. Therefore, the fabrication of all ICs involves the same chain of processes, which are as follows:

1. Wafer Preparation,
2. Epitaxial Growth.
3. Isolation Diffusion.
4. Base Diffusion.
5. Emitter Diffusion.
6. Pre-ohmic Etching and Metallization.

7. Checking and Dicing.
8. Mounting and Packaging.

Wafer Preparation: First of all, a single and pure *p*-type silicon crystal is grown and cut into wafers (thin slices) which are cleaned and polished to a mirror finish. The typical thickness of a finished wafer is 5 mils and its resistivity is 10 Ω-cm. This lightly-doped *p*-type wafer provides the base or 'substrate' on which the circuit components (resistor, transistor; etc.) are to be built.

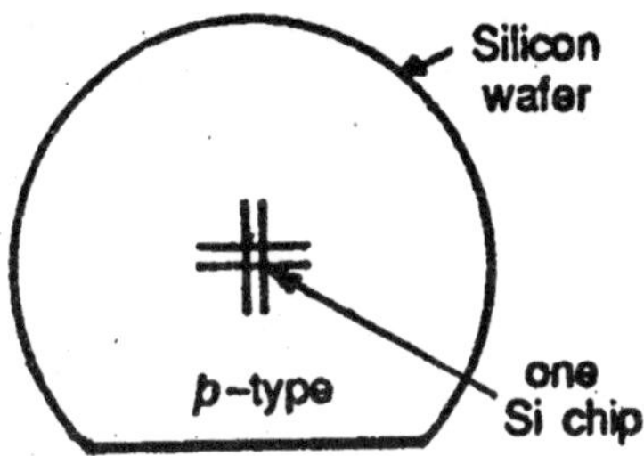

A silicon wafer, about 1.5 inches in diameter, contains several hundreds of chips. In the processes which will follow, exactly identical circuits are produced simultaneously on all the chips. After the final process, the individual chips are separated by cutting.

Epitaxial Growth: The active and passive components are built within a thin *n*-typs 'epitaxial layer' on top. Therefore, an *n*-type layer of silicon, typically 1 mil thick, is grown on the *p* type substrate by placing the wafer in a furnace at 1200°C and introducing as gas containing phosphorus (donor impurity), as shown in Fig. below. The resistivity of this *n*-type layer ranges from. 0 1 to 0 5 .Ω-cm.

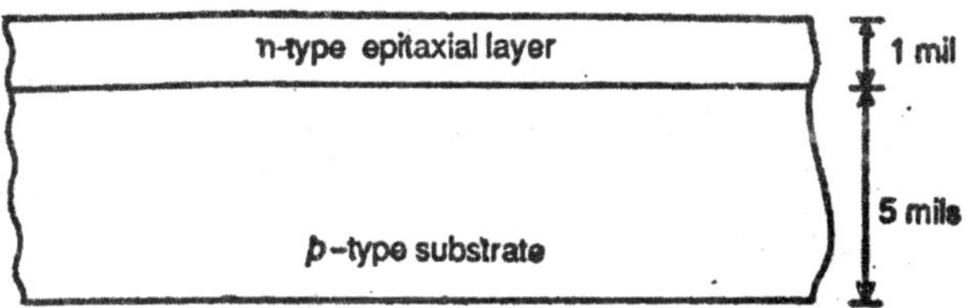

Various Diffusions

Isolation Diffusion: The *n*-type epitaxial layer is isolated into islands, so that each component may be formed on a separate island. This is done by diffusing *p*-type impurity (boron) through the *n*-type epitaxial layer to the *p*-type substrate by a series of steps common in all diffusion processes:

A thin layer (~ 1 micron thick) of silicon dioxide (SiO_2) is formed on the *n*-type epitaxial layer by exposing it to oxygen and heating to about 1000°C (Fig. a). SiO_2 has the property of preventing the diffusion of impurities through it.

Now, in order to form selective openings in SiO_2 through which impurities may be diffused, a photoetching method is used. For this, a thin uniform coating of a photosensitive emulsion, called 'photoresist' is laid on the SiO_2 layer (Fig. b).

Then a large black-and-white layout of the desired pattern of openings is made and reduced photographically. This negative is placed as a "mask" over the photoresist which is then exposed to ultraviolet light (Fig. c).

The photoresist under the transparent regions of the mask becomes polymerized. The mask is now removed, and the wafer is "developed" by a chemical (like trichloroethylene) which dissolves the unexposed (unpolymerized) portions of the photoresist coating and leaves the surface pattern as in Fig. (d).

The wafer is immersed in an etching solution (hydrofluoric acid) which removes SiO_2 from the areas not protected by the photoresist (Fig. e).

The photoresist is removed completely by scrubbing with heated solvents (Fig. f). The wafer is now ready for the isolation diffusion. The remaining SiO_2 serves as a mask for the diffusion of *p*-type impurity (boron).

The wafer is placed in a "boat" and passed through a'

furnace containing boron gas. The *p*-type impurity diffuses into the wafer through the openings in SiO_2, turning the *n*-type material into a p^+–type channel down to a depth extending to the *p*-type substrate.

Thus the *n*-type epitaxial layer is isolated into islands called 'isolation islands' or 'isolated regions' resting on the *p*-type substrate under the SiO_2 layer (Fig. g). Their purpose is to provide electrical isolation between different circuit components The *p*-type substrate is always held at a negative potential with respect to the isolation islands so that the *p-n* junctions are reverse-biased, otherwise the isolation will be lost.

The concentration of the *p*-type impurity atoms in the regions between isolated islands, *i.e.* in p+-type channels, is much higher (and hence indicated as p^+) than that in the *p*-type substrate to prevent any connection between two isolated islands.

The individual circuit components are now built within these isolated islands. Let us consider the fabrication of an *n-p-n* transistor and a resistor in two adjoining islands.

Base Diffusion: A part of the *n*-type island itself provides the collector for the *n-p-n* transistor. The *p*-type base of the transistor is diffused into the collector. At the same time, diffusion of the resistor takes place in the adjoining island. For this a complete new layer of SiO_2 is formed over the wafer and all the above processes are repeated using a different mask so as to create a pattern of openings shown in Fig. below.

The *p*-type impurity (boron) is diffused through these openings. In this way are formed the transistor base region and the resistor. The depth of this diffusion is kept controlled (by controlling the time of diffusion) so that is does not penetrate to the substrate. The resistivity of the base layer is generally much higher than that of the isolation regions.

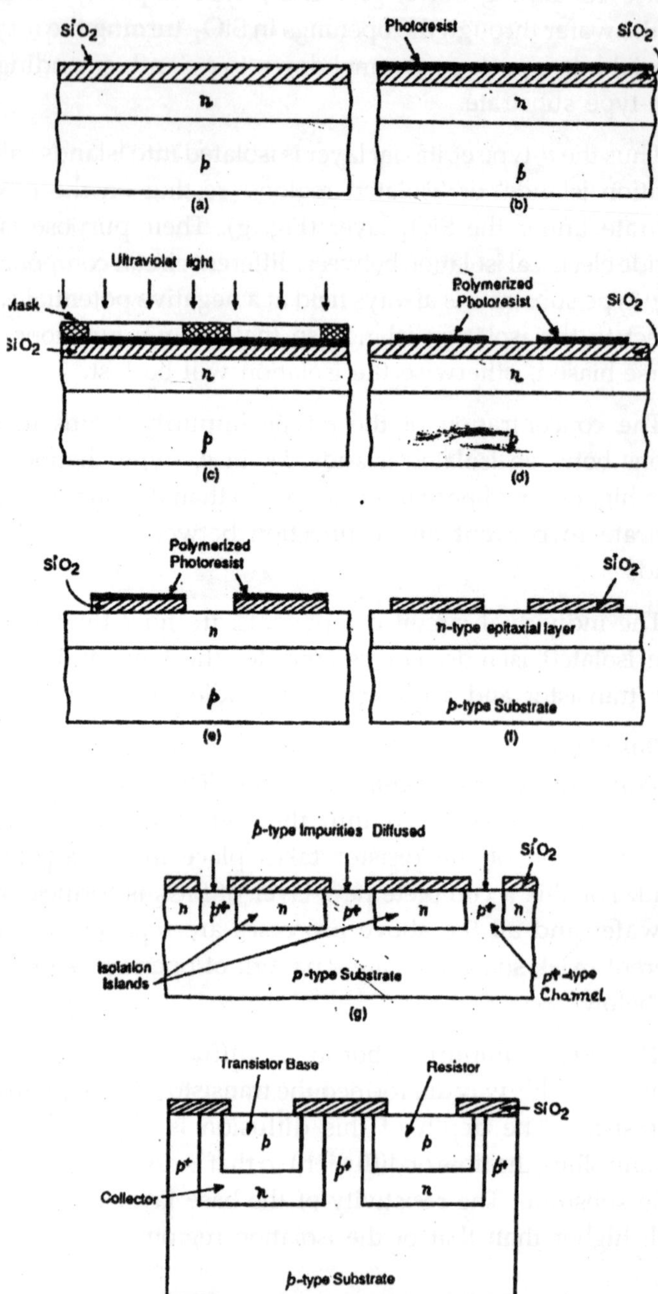
SiO2
n
p
(a)
Photoresist
SiO2
n
p
(b)
Ultraviolet light
Mask
SiO2
n
p
(c)
Polymerized Photoresist
SiO2
n
p
(d)
SiO2
Polymerized Photoresist
n
p
(e)
SiO2
n-type epitaxial layer
p-type Substrate
(f)
p-type Impurities Diffused
SiO2
n
p+
n
p+
n
p+
n
Isolation Islands
p-type Substrate
p+-type Channel
(g)
Transistor Base
Resistor
SiO2
p
p
p+
p+
p+
Collector
n
n
p-type Substrate

Emitter Diffusion: A layer of SiO_2 is again formed over the entire surface and the masking and etching processes are repeated to create an opening in the *p*-type base region, as shown in Fig. below.

Now, *n*-type impurity (phosphorus) of heavy concentration (called n^+) is diffused through this opening for the formation of the transistor emitter.

No diffusion takes place in the adjoining island, since the resistor is complete.

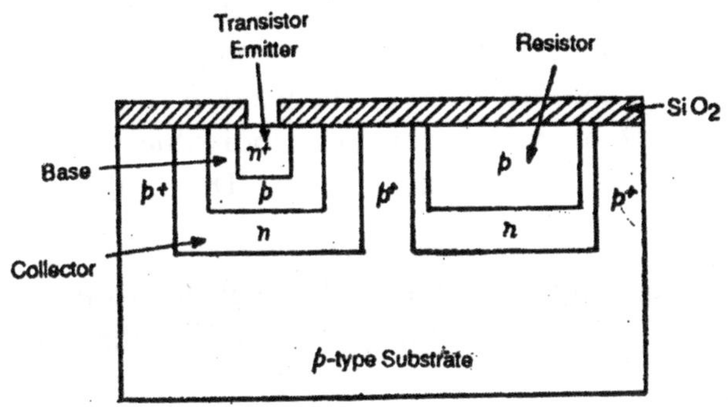

Pre-ohmic Etching and Metallization: In order to make ohmic contacts with the diffused areas, a set of openings is made into a newly formed SiO_2 layer (again doing masking and etching processes) at the points required by the desired circuit.

Inter connections between the various components of the integrated circuit are then made by aluminium metallization. For this a thin coating of aluninium is deposited over the entire wafer by a vacuum evaporation of aluminium, and the undesired aluminium areas are then etched away.

This leaves the desired pattern OF ohmic contacts and interconnection, as shown Fig. below in which an *n-p-n* transistor and a resistor have been fabricated on chip. The circuit symbol is also shown.

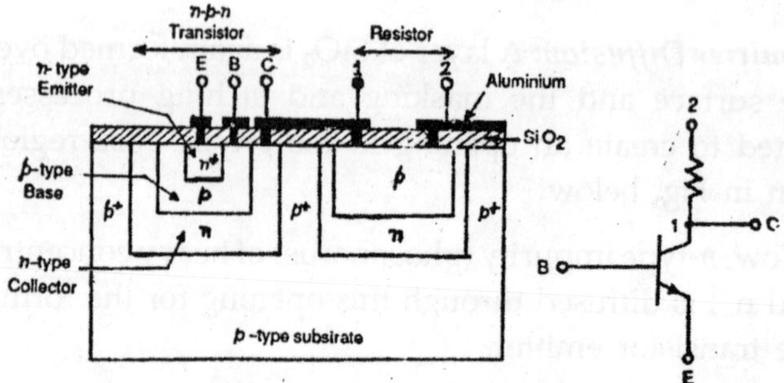

Checking and Dicing: During processing, a large number (several hundred) of ICs are manufactured on a single wafer. After the metallization process has been completed, each IC on the wafer is checked electrically for proper functioning, and the faulty circuits are marked. The wafer is then scribed with a diamond point and separated into individual chips (or dice) containing the integrating circuits. The faulty chips are discarded.

Mounting and Packaging: Individual chips are very small and brittle. Hence each chip is mounted on the gold-plated leads and the chip is provided with 'bonding pads'. Connections between the IC and the package leads are done by aluminium wires from the bonding pads on the chip to the leads on the header. Finally, a cap is placed over the header and the IC is seeded in an inert atmosphere.

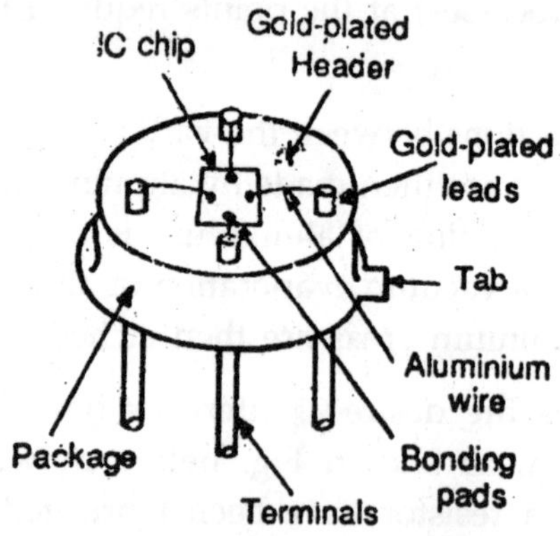

Applications of ICs: Integrated circuits have the advantages of small size, light weight, low cost (because thousands of complex units are fabricated simultaneously), low power consumption, and high reliability. Therefore, they are frequently used in space vehicles, hearing aids and all types of computers.

ICs are of Two Types: linear and non-linear. The linear ICs are used in power amplifiers, high-frequency amplifiers, differential operational amplifiers, voltage regulators and in analog computer circuits.

The non-linear ICs are used in great quantities in digital computers to perform switching functions in logic gates and memory units. The small size of MOS components has led to "large scale integration" (LSI) in which thousands of components are created on a single chip. Such ICs are used, for example, in pocket calculators.

10

Raman Effect

Raman Spectrum

Theory of Raman Effect: When a strong beam of visible or ultraviolet line-spectral light illuminates a gas, a liquid, or a transparent solid, a small fraction of light is scattered in all directions. The spectrum of the scattered light is found to consist of lines of the same frequencies as the incident beam (Rayleigh lines), and also certain weak lines of changed frequencies. These additional lines are called 'Raman lines'.

The Raman lines corresponding to each exciting (Rayleigh) line occur symmetrically on both sides of the exciting line. The lines on the low-frequency side of the exciting line are called 'Stokes' lines, while those on the high-frequency side are called 'anti-Stokes' lines. The anti-Stokes Raman lines are much weaker compared to the Stokes Raman lines. This phenomenon is called 'Raman effect'.

The displacements (in cm^{-1}) of the Raman lines from the corresponding exciting lines are independent of the frequencies of the latter. If another light source with a different line-spectrum is used, other Raman lines are obtained for the same scattering substance. However, the displacements from the exciting lines are the same. For different scattering substances, the displacements have different magnitudes. Thus *the Raman displacements are characteristic of the scattering substance.*

Experimental Setup: The basic requirements for photographing Raman spectrum are a source, a Raman tube and a spectrograph. The source must be an intense line-source with distinct lines in the blue-violet region. A mercury arc or a discharge lamp is a proper source. Now-a-days, laser provides an exceptionally intense and monochromatic Raman source.

The Raman tube used for liquids is a thin-walled glass (or quartz) tube *T* (Fig.) about 15 cm long and 2 cm in diameter, whose one end is closed with an optically-plane glass (or quartz) plate, and the other is drawn out into the shape of a horn and covered with black tape. The flat end serves as the window through which the scattered light emerges, while the blackened horn-shaped end causes the total reflection of the backward scattered light and provides a dark background.

The spectrograph must be one of high light-gathering power combined with good resolution. This may be achieved in a good prism spectrograph with a short-focus camera.

A typical arrangement for obtaining the Raman spectra of liquids is shown in Fig. The source *S* is a long horizontal arc of mercury. The Raman tube *T* containing the experimental liquid is placed above and parallel to the source *S*. In between the source and the tube is placed a glass cylindrical container filled with saturated solution of sodium nitrate. This acts as a cylindrical lens to concentrate the light along the axis of the Raman tube. The sodium nitrate solution absorbs the ultraviolet lines of the mercury arc but transmits the blue line with great intensity. A polished reflector *R* laid over *T* increases the intensity of illumination. The scattered light passing through the plane window of the Raman tube is focussed on the slit of a spectrograph which photographs the spectrum under a long exposure. A spectrophotometer may also be used as a recorder instead of a spectrograph.

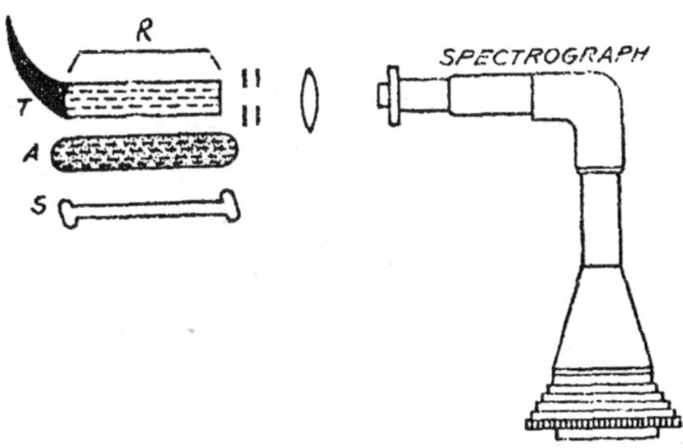

Explanation: Raman effect can be explained from quantum theory. According to this theory, light of frequency v is a bundle of 'photons', each of energy hv. When it falls on a scatterer, the photons collide with the molecules of the scatterer. There are three possibilities in such a collision:

(i) The photon may be scattered or deflected off its path without loss or gain of energy. It then gives rise to the unmodified spectral line of the same frequency v as of the incident light. This is the Rayleigh line.

(ii) The photon may give a part of its energy, ΔE (say), to a molecule which is in its ground energy state E_1 (Fig.). The molecule is then excited to a higher energy state E_2 (= $E_1 + \Delta E$), and the photon is consequently scattered with a smaller energy $hv—\Delta E$. In this case it gives rise to a spectral line of lower frequency (or longer wavelength). This is the Stokes Raman line (Fig.).

(iii) The photon may collide a molecule already in the excited state E_2 and take energy ΔE from it. In this case, the molecule is de-excited to the ground state E_1 and the photon is scattered with increased energy $hv + \Delta E$. Now it gives rise to a spectral line of higher frequency (or shorter wavelength). This is the anti-Stokes Raman line (Fig.).

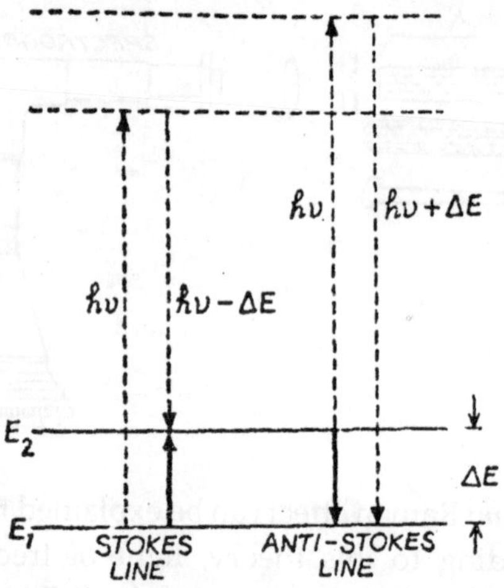

Since the number of molecules in the excited state is very small, the chances of this last process are very small. Hence anti-Stokes Raman lines are much weaker than the Stokes Raman lines.

Uses: Raman effect is a powerful tool for studying the molecular structure of compounds and crystals. It is used to determine the arrangement of atoms in a given molecule. It supplies data regarding the spin and statistics of the nucleus. It is used in industries for studying the composition of mixtures, plastics, etc.

Compton Effect Versus Raman Effect: Both these phenomena are interaction of matter with radiation. In the Compton effect an electron *loosely* bound to an atom acquires energy from the incident photon and becomes free from the atom, the photon being scattered with reduced energy (or reduced frequency). The nucleus takes almost no part in the process.

The Raman effect, on the other hand, is a very general case of interaction between photon and matter in which the entire molecule takes part.

In this process the molecule acquires energy from the incident photon and is simply raised to an excited state of higher energy, no electron from the molecule is freed.

In case of Compton effect the wavelength of the scattered photon is always longer than that of the incident photon. In case of Raman effect, however, the scattered wavelength may be longer as well as shorter than the incident wavelength.

PROBLEMS

1. ***With exciting line 2536 Å a Raman line for a sample is observed at 2612 Å. Calculate the Raman shift in cm^{-1} units.***

Solution: The wave number $\bar{v}\left(=\frac{1}{\lambda}\right)$ of the exciting line is

$$\bar{v} = \frac{1}{2536 \times 10^{-8}\ \text{cm}}$$
$$= 39432\ \text{cm}^{-1}$$

and that of the Raman line is

$$\bar{v}_{\text{Raman}} = \frac{1}{2612 \times 10^{-8}\ \text{cm}}$$
$$= 38285\ \text{cm}^{-1}.$$

∴ Raman shift

$$\Delta\ \bar{v} = 39432 - 38285$$
$$= 1147\ \text{cm}^{-1}.$$

2. ***The exciting line in an experiment is 5460 Å and the Stokes line is at 5520 Å. Find the wavelength of the anti-Stokes line.***

Solution: The Stokes and anti-Stokes lines have the same wave-number displacement with respect to the exciting line. The wave-number of the exciting line is

$$v = \frac{1}{5460 \times 10^{-8} \text{ cm}} = 18315 \text{ cm}^{-1}$$

and that of the Stokes line is

$$\frac{1}{5520 \times 10^{-8} \text{ cm}} = 18116 \text{ cm}^{-1}.$$

Thus the wave-number displacement is

$$\Delta v = 18315–18116 = 199 \text{ cm}^{-1}.$$

Therefore, the wave-number corresponding to the anti-Stokes line would be given by

$$v + \Delta v = 18315 + 199 = 18514 \text{ cm}^{-1}.$$

The corresponding wavelength is

$$\frac{1}{18514 \text{ cm}^{-1}} = 5.401 \times 10^{-5} \text{ cm} = 5401 \text{ Å}.$$

3. *With exciting line 4358 Å, a sample gives Stokes line at 4458 A. Deduce the wavelength of the anti-Stokes line.*

[**Ans.** 4262 Å]

Zeeman Effect

Pattern of Vector

Orbital Magnetic Dipole Moment of Atomic Electron: An electron revolving in an orbit about the nucleus of an atom is a minute current loop and produces a magnetic field. It thus behaves like a magnetic dipole. Let us calculate its magnetic moment. Let us consider an electron of mass m and charge $-e$ moving with speed v in a circular Bohr orbit of radius r (Fig.), It constitutes a current of magnitude

$$i = \frac{e}{T},$$

where T is the orbital period of the electron.

Now, $$T = \frac{2\pi r}{v}, \text{ and so}$$

$$i = \frac{ev}{2\pi r}.$$

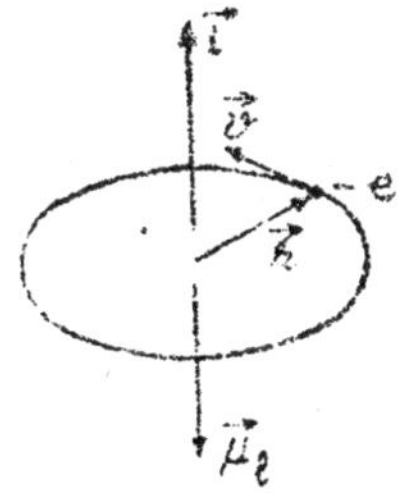

From electromagnetic theory, the magnitude of the orbital magnetic dipole moment $\vec{\mu}_l$ for a current i in a loop of area A is

$$\mu_l = iA$$

and its direction is perpendicular to the plane of the orbit as shown. Substituting the value of i from above and taking $A = \pi r^2$, we have

$$\mu_l = \frac{ev}{2\pi r}\pi r^2 = \frac{evr}{2} \qquad \text{... } (i)$$

Because the electron has a negative charge, its magnetic dipole moment $\vec{\mu}_l$ is opposite in direction to its orbital angular momentum $\vec{L}$, whose magnitude is given by

$$L = mvr. \qquad \text{... } (ii)$$

Dividing eq. (*i*) by eq. (*ii*), we get

$$\frac{\mu_l}{L} = \frac{e}{2m} \qquad \text{... } (iii)$$

Thus the ratio of the magnitude μ_l of the orbital magnetic dipole moment to the magnitude L of the orbital angular momentum for the electron is a constant, independent of the details of the orbit. This constant is called the 'gyromagnetic ratio' for the electron.

We can write eq. (*iii*) as a vector equation:

$$\boxed{\vec{\mu}_l = -\left(\frac{e}{2m}\right)\vec{L}.}$$

The minus sign means that $\vec{\mu}_l$ is in the opposite direction to $\vec{L}$.

The unit of electron magnetic moment is amp-m^2 or joule/tesla.

Bohr Magneton: From wave mechanics, the permitted scalar values of $\vec{L}$ are given by

$$L = \sqrt{[l(l+1)]}\,\frac{h}{2\pi},$$

where l is the 'orbital quantum number'. Therefore, the magnitude of the orbital magnetic moment of the election is

$$\mu_l = \sqrt{[l(l+1)]}\,\frac{eh}{4\pi m}.$$

The quantity $\frac{eh}{4\pi m}$ forms a natural unit for the measurement of atomic magnetic dipole moments, and is called the 'Bohr magneton', denoted by μ_{B}. Thus

$$\mu_l = \sqrt{[l(l+1)]}\,\mu_B,$$

where $$\mu_B = \frac{eh}{4\pi m} = \frac{(1.6\times10^{-19}\,C)(6.63\times10^{-34}\text{ J-s})}{4\times3.14(9.1\times10^{31}\text{kg})}$$

$$= 9.28 \times 10^{-24}\text{ amp-m}^2.$$

Larmor Precession (Behaviour of a Magnetic Dipole in an External Magnetic Field): An electron moving around the nucleus of an atom is equivalent to a magnetic dipole. Hence, when the atom is placed in an external magnetic field, the electron orbit precesses about the field direction as axis. This precession is called the 'Larmor precession, and the frequency of this precession is called the 'Larmor frequency'.

In Fig. is shown an electron orbit in an external magnetic field $\vec{B}$. The orbital angular momentum of the electron is represented by a vector $\vec{L}$ perpendicular to the plane of the orbit. Let θ be the angle between $\vec{L}$ and $\vec{B}$.

The orbital dipole moment $\vec{\mu}_l$ of the electron is given by

$$\vec{\mu}_l = -\left(\frac{e}{2m}\right)\vec{L}, \qquad \text{... (i)}$$

where $-e$ is the charge on the electron of mass m. The minus sign signifies that $\vec{\mu}_l$ is directed opposite to $\vec{L}$. As a result of its interaction with $\vec{B}$, the dipole experiences a torque $\vec{\tau}$, given by

$$\vec{\tau} = \vec{\mu}_l \times \vec{B}. \qquad \text{... (ii)}$$

According to eq. (*i*) and (*ii*), the torque $\vec{\tau}$ is always *perpendicular* to the angular momentum $\vec{L}$.

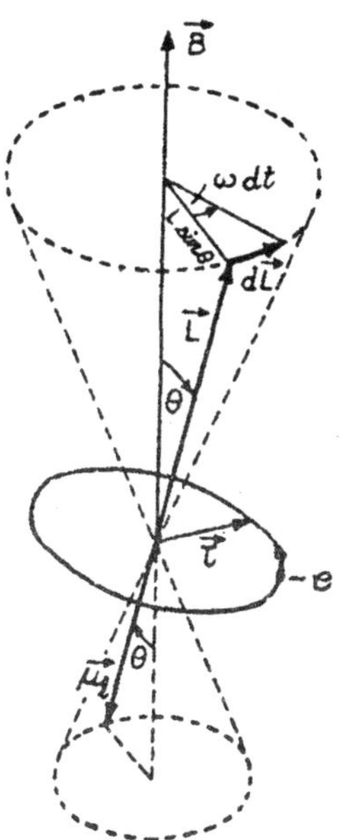

We know that a torque causes the angular momentum to change according to a form of Newton's law

$$\vec{\tau} = \frac{d\vec{L}}{dt},$$

and the change takes place in the direction of the torque. The torque $\vec{\tau}$ on the electron, therefore, produces a changed $\vec{L}$ in $\vec{L}$ in a time *dt*. *The changed* $\vec{L}$ is *perpendicular to* $\vec{L}$ (because the change is in the direction of torque, and the torque is perpendicular to $\vec{L}$). Hence the angular momentum $\vec{L}$ remains constant in magnitude, but its direction changes. As time goes on, $\vec{L}$ traces a cone around $\vec{B}$, such that the angle between $\vec{L}$ and $\vec{B}$ remains constant. This is the precession of $\vec{L}$, and hence of the electron orbit, around $\vec{B}$.

If ω be the angular velocity of precession, then $\vec{L}$ precesses through an angle ω *dt* in time *dt*. From the Fig., we see that

$$\omega\, dt = \frac{dL}{L\sin\theta} \qquad \left[\text{angle} = \frac{\text{arc}}{\text{radius}}\right]$$

or

$$\omega = \frac{dL}{dt}\frac{1}{L\sin\theta} = \frac{\tau}{L\sin\theta}.$$

But. from eq. (*ii*), $\tau = \mu_l B \sin\theta$.

$$\therefore \qquad \omega = \frac{\mu_l}{L}B$$

Thus the angular velocity of Larmor precession is equal to the product of the magnitude of the magnetic field and the ratio of the magnitude of the magnetic moment to the magnitude of the angular momentum.

Again, from eq. (*i*), $\dfrac{\mu_l}{L} = \dfrac{e}{2m}$.

$$\therefore \qquad \omega = \frac{e}{2m} B.$$

The Larmor frequency (frequency of precession) is therefore

$$f = \frac{\omega}{2\pi} = \frac{e}{4\pi m} B.$$

It is independent of the orientation angle θ between orbit normal ($\vec{L}$) and field direction ($\vec{B}$).

Importance: This theorem is of considerable importance in atomic structure as it enables an easy calculation of energy levels in the presence of an external magnetic field.

Nr: — Putting $e = 1.6 \times 10^{-19}$ coul, $m = 9.1 \times 10^{-31}$ kg and $B - 10^4$ weber/m^2 (given), we get

$$f = 1.4 \times 10^{14} \text{ per second.}$$

Space Quantisation of Atoms: An electron moving around the nucleus of an atom is equivalent to a magnetic dipole. Hence, when the atom is placed in an external magnetic field $\vec{B}$, the electron orbit precesses about the field direction as axis. The electron orbital angular momentum vector $\vec{L}$ traces a cone around $\vec{B}$ such that the angle θ between $\vec{L}$ and $\vec{B}$ remains constant (Fig.).

If the magnetic field $\vec{B}$ is along the z-axis, the component of $\vec{L}$ parallel to the field is

$$L_z = L \cos \theta$$

or
$$\cos \theta = \frac{L_z}{L}$$

Quantum mechanically, the angular momentum L and its z-component L_z are quantised according to the relations

$$L = \sqrt{[l(l+1)]} \frac{h}{2\pi}$$

and $$L_z = m_l \frac{h}{2\pi},$$

where l and m_l are the orbital and magnetic quantum numbers respectively. Therefore,

$$\cos\theta = \frac{L_z}{L} = \frac{m_l}{\sqrt{[l(l+1)]}}.$$

Thus the angle θ between $\vec{L}$ and the z-axis is determined by the quantum numbers l and m_l. Since, for a given l, there are $(2l+1)$ possible values of ml (= 0, ± 1, ± 2,........., ± 1), the angle θ can assume $(2l + 1)$ discrete values. In other words, the angular momentum vector $\vec{L}$ can have $(2l+1)$ discrete orientations with respect to the magnetic field. This quantisation of the orientation of atoms in space is known as 'space quantisation'.

The space quantisation of the orbital angular momentum vector $\vec{L}$ corresponding to $l = 2$ {or $L = \sqrt{(6)}\, h/2\pi$} is shown in Fig. For $l = 2$, we have

$$ml = 2, 1, 0, -1, -2$$

so that $$L_z = 2\frac{h}{2\pi}, \frac{h}{2\pi}, 0, -\frac{h}{2\pi}, -2\frac{h}{2\pi}.$$

Alternatively, the orientations θ of $\vec{L}$ with respect to the field $\vec{B}$ (*z-axis*) are given by

$$\cos\theta = \frac{m_l}{\sqrt{[l(l+1)]}}$$

$$= \frac{2}{\sqrt{(6)}}, \frac{1}{\sqrt{(6)}}, 0, -\frac{1}{\sqrt{(6)}}, -\frac{2}{\sqrt{(6)}},$$

$$= 0.8165, 0.4082, 0, -0.4082, -0.8165$$

or $$\theta = 35^\circ, 66^\circ, 90^\circ, 114^\circ, 145^\circ$$

We note that $\vec{L}$ can never be aligned exactly parallel or antiparallel to $\vec{B}$, since $|m_l|$ is always smaller than $\sqrt{[l(l+1)]}$.

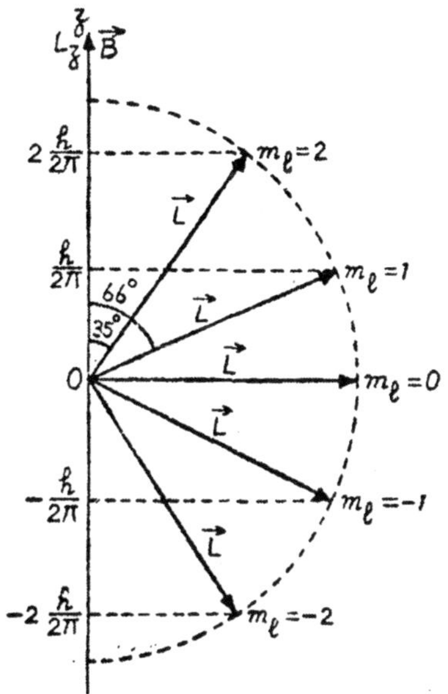

Effect of Zeeman

Zeeman, in 1896, observed that when a light-source giving line spectrum is placed in an *external* magnetic field, the spectral lines emitted by the atoms of the source are split into a number of polarised components. This effect of magnetic field on the atomic spectral lines is called 'Zeeman effect'.

To produce Zeeman effect, a source of light, such as a gas discharge tube, is placed symmetrically between the pole-pieces of a strong electromagnet, one of whose pieces carries a hole drilled parallel to the magnetic field direction. The light coming from the tube is examined by a spectroscope of high resolving power. The attention is focused on a single spectral line observable in the absence of the magnetic field (Fig.).

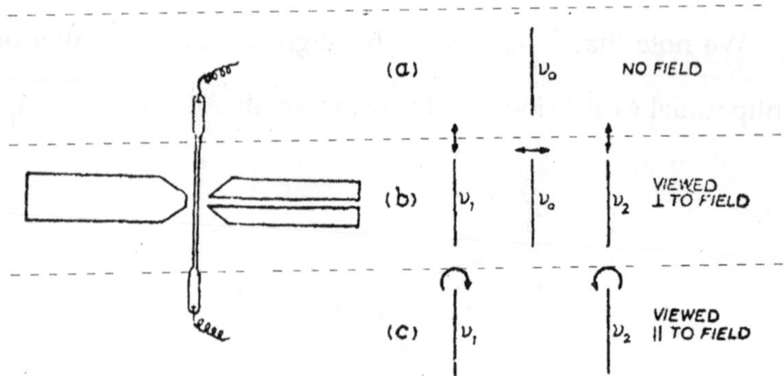

When the light is viewed at right angles to the magnetic field direction, a singlet spectral line is found to split up into three components (Fig.). The central component is in the same position and hence has the same frequency v_0 as the original line. The outer components of frequencies v_1 and v_2 are displaced equally from the central component. The central component is linearly polarised with electric vector vibrating parallel to the magnetic field, while the two outer components are linearly polarised with electric vector vibrating at right angles to the field. This splitting is known as '*normal* transverse Zeeman effect'.

When the light is viewed (through the hole) in a direction parallel to the magnetic field, only the two outer components are seen and there is no central component (Fig.). These components are circularly polarised in opposite senses. This is known as '*normal* longitudinal Zeeman effect'.

The fine-structure components of a *multiplet* spectral line, however, show a complex Zeeman pattern. For example, the D_1 and D_2 components of sodium yellow doublet give four and six lines respectively in the Zeeman pattern. This is 'anomalous' Zeeman effect.

Explanation of Normal Zeeman Effect: The normal Zeeman effect, which is shown by spectral lines arising from transitions between the *singlet* energy levels of an atom, can be explained from the classical electron theory and also from the quantum theory (without taking note of electron spin).

In terms of quantum theory, an atom possesses an electron orbital angular momentum $\vec{L}$ and an orbital magnetic moment $\vec{\mu}_l$, with gyromagnetic ratio given by

$$\frac{\mu_l}{L} = \frac{e}{2m},$$

where e and m are the charge and mass of electron. The vector $\vec{\mu}_l$ is directed opposite to the vector $\vec{L}$ because the electron is negatively charged (Fig.).

When the atom is placed in an external magnetic field $\vec{B}$ say along the z axis, the vector $\vec{L}$ precesses around the field direction (Larmor precession) with quantised components given by

$$L_z = m_l \frac{h}{2\pi},$$

where m_l is the magnetic quantum number.

By Larmor's theorem, the angular velocity of precession is given by

$$\omega = \frac{\mu_l}{L} B = \frac{e}{2m} B.$$

The change in energy of such a precession is equal to the product of the angular velocity and the component of angular momentum along the field, that is,

$$\Delta E = \omega L_z = \frac{eB}{2m} mt \frac{h}{2\pi} = \text{mt} \frac{ehB}{4\pi m}. \quad \dots (i)$$

Now, *ml* takes discrete values from + *l* to —*l*, *i.e.*

ml =*l*, *l*–1,...2, 1,0, – 1, –2, ...*l*, a total of (2*l*+1) values. Thus each energy level (of a given *l* value) of the atom placed in the magnetic field is split into (2*l* + 1) energy levels, called 'Zeeman levels', having different m_l, values. The separation between the Zeeman levels is same for all the energy levels of the atom.

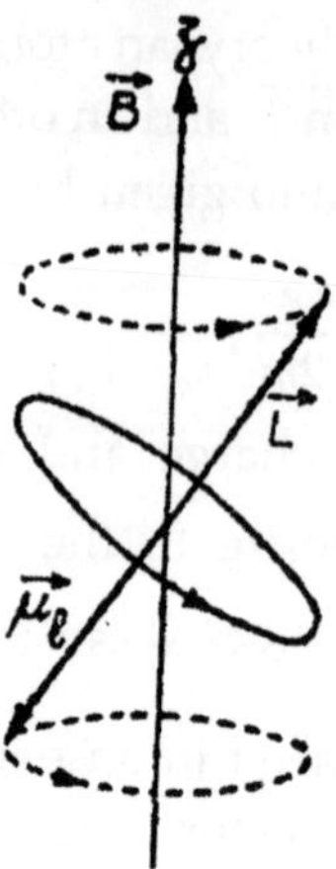

Let us consider a line arising by electron transition from $l = 1$ level to $l = 0$ level (Fig.). In the magnetic field, the level $l = 1$ is split into $(2l + 1) = 3$ components corresponding to $ml = +1, 0, -1$; while the level $l = 0$ remains unplatted (Fig.).

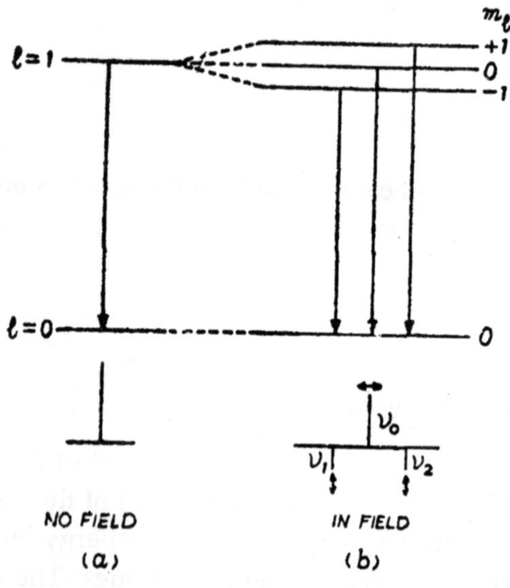

Theory shows that only those transitions are allowed for which the quantum number m_l changes by 0 or ± 1, *i.e.*

$$\Delta m_l = 0 \pm 1.$$

Therefore, from eq. (*i*), the energy-change due to the Zeeman splitting of levels is

$$\Delta E = 0, \pm \frac{eh}{4\pi m} B.$$

Hence the change in the frequency of the emitted spectral line is given by

$$\Delta v = \frac{\Delta E}{h} = 0 \pm \frac{eB}{4\pi m}.$$

Thus the original line of frequency v_0 is splitted into three components of frequencies

$$v_1 = v_0 + \frac{{}^{v_0,}\, eB}{4\pi m},$$

and $$v_2 = v_0 - \frac{eB}{4\pi m},$$

i.e. one displaced component v_0, and two equally-displaced components on either side of the v_0 component (Fig.).

When the light is viewed parallel to the magnetic field B, the displaced components v_1 and v_2 (which are circularly-polarised) are seen. The undisplaced component, however, which has optical vibrations parallel to the field does not send light in this direction due to the *transverse* nature of light waves. Hence in this case no central component is seen.

Determination of e/m: The change in the frequency of a spectral line, when the light source is placed in a magnetic field B, is given by

$$\Delta v = \frac{eB}{4\pi m}$$

If λ be the wavelength of the spectral line, then

$$v = \frac{c}{\lambda}$$

or $$\Delta v = -\frac{c}{\lambda^2}\Delta\lambda$$

$\therefore$ $$\Delta\lambda = -\frac{\lambda^2}{c}\Delta v = -\frac{\lambda^2}{c}\frac{eB}{4\pi m}$$

or $$\Delta\lambda = \frac{eB\lambda^2}{4\pi mc} \text{ (numerically).}$$

$\therefore$ $$\frac{e}{m} = \frac{4\pi c}{B}\left(\frac{\Delta\lambda}{\lambda^2}\right).$$

Hence, measuring the wavelength-change, e/m can be evaluated.

Spin of Electron

The Bohr-Sommerfeld quantum theory of elliptic orbits *with relativity correction* was in fair agreement with the observed fine structure of hydrogen spectral lines. It, however, suffered from two major drawbacks:

Firstly, it could not explain the fine-structure observed in the spectral lines of atoms other than hydrogen. For example, the spectral lines of alkali atoms are *doublets*, have two close fine-structure components. In alkali atoms, the (single) optical electron moves in a Bohr-like orbit of *large* radius at *low* velocity. Therefore, the relativity effect would be too small to account for the large fine-structure splitting observed in alkali lines.

Secondly, the simple quantum theory failed to explain anomalous Zeeman effect, that is, the splitting of atomic spectral lines into four, six, or more, components when the light source was placed in an external magnetic field.

In view of these drawbacks of the theory, Goudsmit and Uhlenbeck proposed in 1925 that *an electron must be looked upon as a charged sphere spinning about its own axis having an intrinsic angular momentum and an intrinsic magnetic moment.* These are called 'spin angular momentum' $\vec{S}$ and 'spin magnetic moment' $\vec{\mu}_s$, respectively. (These are besides the orbital angular momentum $\vec{L}$ and orbital magnetic moment $\vec{\mu}_l$). The magnitude of the spin angular momentum is

$$S = \sqrt{[s(s+1)]}\frac{h}{2\pi},$$

where s is the 'spin quantum number'. The only value s can have is

$$S = \frac{1}{2},$$

as required by experimental data. Thus

$$S = \sqrt{\left[\frac{1}{2}\left(\frac{1}{2}+1\right)\right]}\frac{n}{2\pi} = \frac{\sqrt{3}}{2}\frac{h}{2\pi}.$$

The component of $\vec{S}$ along a magnetic field parallel to the z-direction is

$$S_z = m_s \frac{h}{2\pi},$$

where m_s is the 'spin magnetic quantum number' and takes $(2s + 1) = 2$ values which are $+ s$ and $- s$, that is

$$m_s = +\frac{1}{2} \text{ and } -\frac{1}{2}.$$

Thus $$s_z = +\frac{1}{2}\frac{h}{2\pi} \text{ and } = -\frac{1}{2}\frac{h}{2\pi}$$

The gyromagnetic ratio for electron spin, $\frac{\mu_s}{S}$ is *twice* the corresponding ratio $\frac{\mu_l}{L}\left(=\frac{e}{2m}\right)$ for the electron orbital motion.

Thus the spin magnetic moment $\vec{\mu}_s$ of electron is related to the spin angular momentum $\vec{S}$ by

$$\vec{\mu}_s = -\frac{e}{m}\vec{S}.$$

The minus sign indicates that $\vec{\mu}_s$ is opposite in direction to $\vec{S}$ (because electron is negatively charged).

The magnitude of the spin magnetic moment is

$$\mu_s = \frac{e}{m}S.$$

$$= \frac{e}{m}\frac{\sqrt{3}}{2}\frac{h}{2\pi}$$

$$= \sqrt{3}\frac{eh}{4\pi m}$$

$$= \sqrt{3}\,\mu_{B,}$$

where μ_B is the Bohr magneton.

Coupling of Orbital and Spin Angular Momenta: Vector Model of the Atom: The total angular momentum of an atom results from the combination of the orbital and spin angular momenta of its electrons. Since angular momentum is a vector quantity, we can represent the total angular momentum by means of a vector, obtained by the addition of orbital and spin angular momentum vectors. This leads to the vector model of the atom.

Let us consider an atom whose total angular momentum is provided by a single electron. The magnitude of the orbital angular momentum $\vec{L}$ of an atomic electron is given by

$$L = \sqrt{[l(l+1)]}\,\frac{h}{2\pi}$$

and its z component

$$Lz = m_l\frac{h}{2\pi},$$

where l is orbital quantum number and ml, is the corresponding magnetic quantum number, with values

$$m_l = l, l - 1, \ldots\ldots\ldots, -l + 1, -l.$$

Similarly, the magnitude of the spin angular momentum $\vec{S}$ is given by

$$S = \sqrt{s(s+1)}\frac{h}{2\pi}$$

and its z-component

$$S_z = m_s \frac{h}{2\pi},$$

where s is the spin quantum number (which has the sole value $+\frac{1}{2}$) and m_s is magnetic spin quantum number ($m_s = \pm\frac{1}{2} = \pm s$).

The total angular momentum of the one-electron atom, $\vec{J}$, is the vector sum of $\vec{L}$ and $\vec{S}$, that is

$$\vec{J} = \vec{L} + \vec{S}.$$

The magnitude and the z-component of $\vec{J}$ are specified by two quantum numbers j and m_j, according to the usual quantisation conditions

$$J = \sqrt{j(j+1)}\frac{h}{2\pi}$$

and $$J_z = mj \frac{h}{2\pi}.$$

j is called the 'inner quantum number' and m_j is the corresponding magnetic quantum number. The possible values of m, range from $+j$ to $-j$ in integral steps:

$$m_j = j, j - 1 \ldots\ldots\ldots, -j + 1, -j.$$

Let us obtain the relationship among the various angular momentum quantum numbers. Since J_z, L_x and S_z are scalar quantities, we may write

$$J_z = L_z \pm S_z.$$

This gives $\quad m_j = mi \pm m_s.$

The maximum values of m_1, m_l and m_s are j, l and s respectively. Therefore, we have

$$\boxed{j = l \pm s}$$

Since $\vec{J}$, $\vec{L}$ and $\vec{S}$ are ail quantised, they can have only certain specific relative orientations. In case of a one-electron atom, there are only two relative orientations possible, corresponding to

$$j = l + s,$$

so that $\quad J > L.$

and $\quad j = l - s,$

so that $\quad J < L.$

The two ways in which $\vec{L}$ and $\vec{S}$ can combine to form $\vec{J}$ (when $l = 1$, $s = \frac{1}{2}$) are shown in Fig.

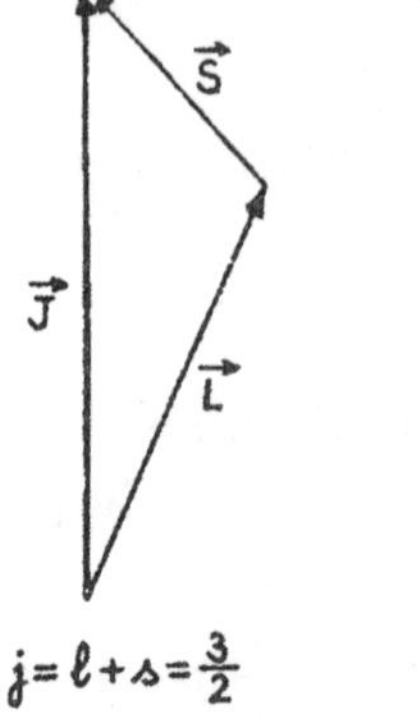

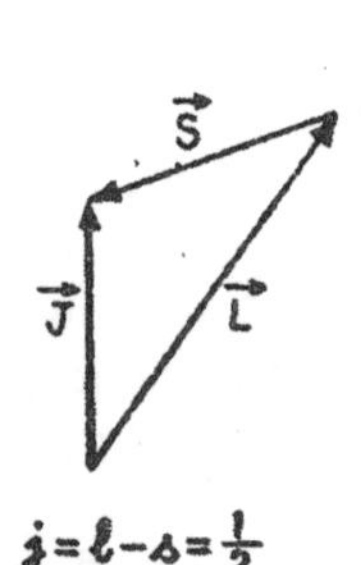

The angular momenta of the atomic electron, $\vec{L}$ and $\vec{S}$, interact magnetically; which is known as 'spin-orbit interaction'. They exert torques on each other. These *internal* torques do not change the magnitudes of the vectors $\vec{L}$ and $\vec{S}$, but cause them to precess uniformly around their resultant $\vec{J}$ (Fig.). If the atom is in free space so that no external torques act on it, then the total angular momentum $\vec{J}$ is conserved in magnitude and direction. Obviously, the angle between $\vec{L}$ and $\vec{S}$ would remain invariant. This is the vector model of the one-electron atom. It can be extended to many-electron atoms. The vector model enables us to explain the phenomena which could not be understood from Bohr-Sommerfeld theory such as fine structure of spectral lines and anomalous Zeeman effect.

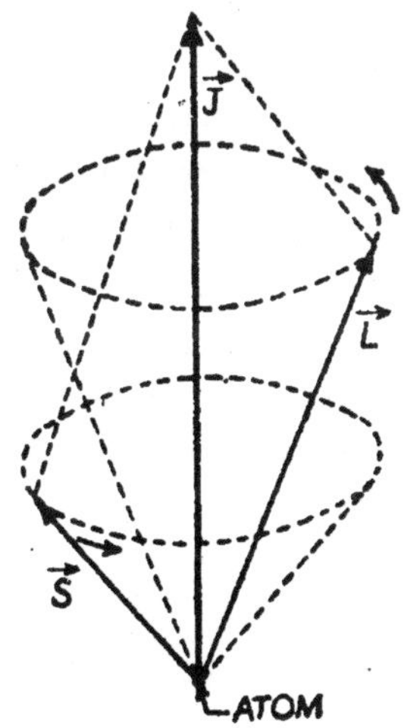

Alkali Spectra: The alkali atoms Li, Na, K,.........readily give up one electron to form positive ions. The energy required to remove one electron from these atoms is small (5.1 eV in case of Na), but that required to remove a second electron is much larger (47.3 eV). This suggests that of all the electrons in an alkali atom, *one* electron is loosely-bound to the atom. The spectral lines of an alkali atom arise due to the transitions of this electron only which is called the 'optical' or 'valence' electron. The alkali spectrum is called 'one-electron' spectrum.

The alkali spectrum consists of spectral lines which can be classified into four series: principal, series, sharp series, diffuse series and fundamental series. The principal series is the most prominent and can be observed in emission as well as in absorption spectrum. The other series are observed in emission spectrum only.

The emission of alkali spectral lines can be fairly explained on the same lines as the Bohr-Sommerfeld theory for hydrogen atom. An atom has a number of discrete energy states, each state being characterised by a total quantum number $n(= 1, 2, 3, \ldots \infty)$. For each value of n, there are component levels labelled by an additional quantum number l, called the 'orbital' quantum number, l can take values 0, 1, 2,......$(n - 1)$. Thus $n = 1$ state has only one level ($l = 0$); $n = 2$ state has two levels ($l = 0, 1$), and so on. The levels corresponding to $l = 0, 1, 2, 3, \ldots$ are called as *s, p, d, f*,....levels respectively. Thus $n - 1$ state has a level called 1s; the $n = 2$ state has two levels called 2*s* and 2*p*; the $n = 3$ state has three levels called 3*s*, 3*p* and 3*d*; the $n = 4$ state has four levels 4*s*, 4*p*, 4*d*, 4*f*, and so on. The energies of these levels are given by

$$E_{n,l} = -\frac{Rhc}{(n - \Delta)^2},$$

where Δ is called 'quantum defect' and depends on l. Thus the energies of levels with same n but different l are *different*.

Let us now consider an alkali atom, say sodium (Na). Measurement of spectral wavelengths and the ionisation potential shows that the ground state of the sodium atom is 3*s*, *i.e.* the optical electron occupies the 3*s* level in the normal atom. When the atom is excited by some outer means, the electron leaves the 3*s* level and goes to *any* of the higher levels 3*p*, 3*d*, 4*s*, 4*p*, 4*d*, 4*f*,.... From the higher level the electron jumps back to the lower levels but only such that the l value changes by ± 1 (Fig.), *i.e.*

$$\Delta l = \pm 1.$$

This is called the 'selection rule'. When the electron jumps from any p-level to the lowest s-level (3s), it emits a line of principal series; when it jumps from any s-level to the lowest p-level (3p), it emits a line of sharp series; and so on. Thus emission of all the spectral series is explained.

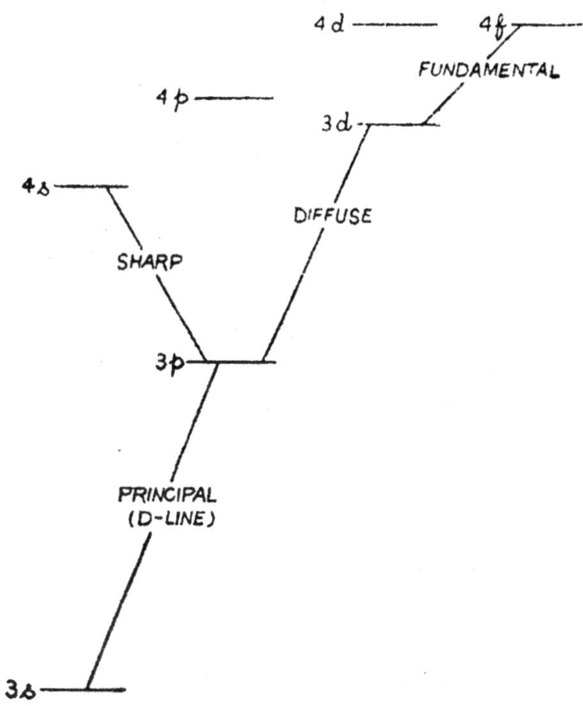

Fine Structure of Alkali Spectra and its Explanation: When the spectral lines of an alkali atom are seen under high resolution, each of them is found to consist of two close components. For example, the yellow D-line of sodium consists of two close lines of wavelengths 5890 Å and 5896 Å. This is called the fine structure and is explained by introducing the conception of electron spin.

The total angular momentum of the electron is the *vector* sum of its orbital angular momentum $\sqrt{[l(l+1)]}\,\frac{h}{2\pi}$ and the spin angular momentum $\sqrt{[s(s+1)]}\,\frac{h}{2\pi}$ and is given by $\sqrt{[j(j+1)]}\,\frac{h}{2\pi}$, where j is called the 'inner quantum number'. It takes the values given by

$$j = l \pm s = l \pm \frac{1}{2}.$$

Thus, for

$$l = 0 \text{ (s-level)}, j = \frac{1}{2} \text{ (j cannot be negative)}$$

$$l = 1 \text{ (p-level)}, j = \frac{3}{2}, \frac{1}{2}$$

$$l = 2 \text{ (d-level)}, j = \frac{5}{2}, \frac{3}{2}, \text{ and so on.}$$

Each value of the total angular momentum of the electron corresponds to a particular total energy of the electron. Therefore, each energy level of a given l-value is split into two sub-levels of slightly different energies, corresponding to the two j-values. The s-levels (for which $l = 0$) still remain unsplitted because there is only one j-value for them. On this basis, the doublet structure of the spectral lines of alkali atoms can be explained.

Let us consider the sodium D-line. It arises from the transition of the electron from the $3p$ to the $3s$ level (Fig.). But, due to electron spin, the $3p$ level consists of two sub-levels, one corresponding to $j = \frac{3}{2}$ and the other corresponding *to* $j = \frac{1}{2}$ (Fig.). Thus there are two transitions and hence the D-line is a doublet.

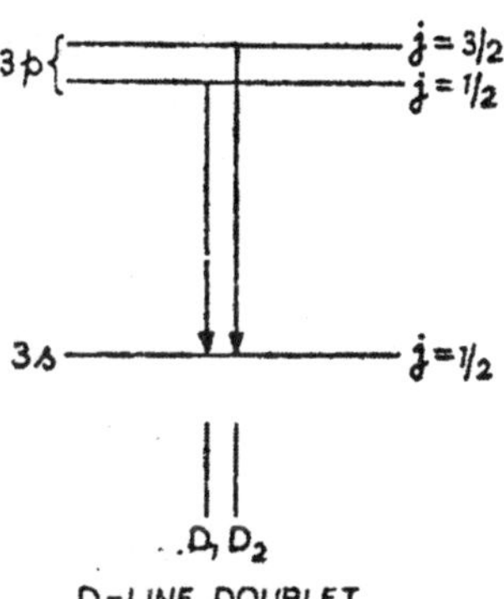

Similarly, the spin-orbit coupling together with relativity correction explains the fine structure of hydrogen lines.

Stern-Gerlach Experiment: Stern and Gerlach, in 1921, performed an experiment which demonstrated directly that an atom placed in a magnetic field can take only certain discrete orientations with respect to the field (space quantisation). It also demonstrated the existence of electron spin, thus providing an experimental verification of the vector model of atom.

The plan of the experiment is shown in Fig. A beam of neutral silver atoms was formed by heating silver in an oven. It was collimated by a few fine slits and then passed through a *non-homogeneous* magnetic field. The field was produced by specially-designed pole-pieces whose cross-sectional view is displayed separately. It shows that the field increases in intensity in the z-direction defined in the figure.

The beam leaving the magnetic field was received on a photographic plate *P*. On developing the plate, no trace of the direct beam was obtained. Instead, two traces were obtained, symmetrically situated with respect to the direct beam. This meant that the beam of silver atoms splitted into two discrete components, one component being bent in the + z-direction and the other bent in the — z-direction. The experiment was repeated using other atoms, and in each case the beam was found splitted into two, or more, discrete components. This result is interpreted in the following way:

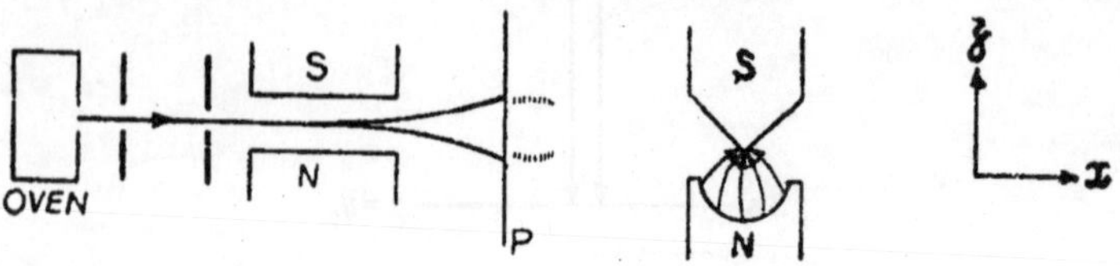

A magnet experiences a net *deflecting* force in a non-homogeneous magnetic field which depends on the orientation of the magnet in the field. Since atoms are tiny magnets, they experience deflecting force when passing through the field. If an atom could have *any* orientation in the magnetic field, then for the millions of atoms present in the beam, all possible orientations would be obtained and the beam would be deflected into a continuous band.

In the experiment, however, there was no band, but discrete traces on the photographic plate. This showed that *the atoms passing through the field were oriented in space in discrete directions* so that the beam deflected in certain discrete directions only and gave discrete traces on the plate.

The experiment is also an evidence for the existence of electron spin. This was shown most clearly in 1927 by Phipps and Taylor, who repeated the Stern Gerlach experiment by using a beam of hydrogen atoms. This atom consists of a single electron which, in the ground state, lies in an *s*-level, for which the quantum number $l = 0$. If there were no spin, then j would also be zero, so that $(2j +.1) = 1$.

In that case the hydrogen atomic beam would be unaffected by the magnetic field, and only one trace would be obtained on the plate. Phipps and Taylor, however, found the beam to be splitted into two symmetrically deflected components giving rise to two traces. This is just the case when the existence of electron spin is admitted and a value $\frac{1}{2}$ is assigned to the spin quantum number. Thus $j = l \pm s = 0 \pm \frac{1}{2} = \frac{1}{2}$ so that $2j + 1 = 2$.

Hence two traces.

Need of Inhomogeneous Magnetic Field: If the field were homogeneous, then the atoms (tiny magnets) would have experienced only a turning moment, and no deflecting force. As such, we could not obtain the deflected components in spite of the orientation of the atoms relative to the field.

Atoms, not Ions: In the Stern-Gerlach experiment, a beam of `neutral' atoms is passed through an inhomogeneous magnetic field, and each atom experiences a transverse force depending upon its orientation with respect to the field. If (charged) ions were used, they would be subjected to Lorentz force also and their deflection would no longer be transverse so that no traces would be obtained on the plate.

PROBLEMS

1. *Calculate the two possible orientations of spin vector $\vec{S}$ with respect to magnetic field direction.*

Solution: Let the magnetic field $\vec{B}$ be along the z-axis. The magnitude of the spin angular momentum $\vec{S}$ and its z-component are quantised according to the relations

$$S = \sqrt{[s(s+1)]}\,\frac{h}{2\pi}, \qquad s = \frac{1}{2}$$

and

$$S_z = m_s\,\frac{h}{2\pi}, \qquad m_s = \pm\frac{1}{2}$$

Hence the angle θ between $\vec{S}$ and the z-axis (Fig.) is determined by the quantum numbers m_s *and* s, according as

$$\cos\theta = \frac{S_z}{S} = \frac{m_s}{\sqrt{[s(s+1)]}}$$

$$= \frac{2}{\sqrt{3}}\, m_s. \qquad \left[\because s = \frac{1}{2}\right]$$

For $\quad m_s = +\frac{1}{2}$, we have

$$\cos\theta = +\frac{1}{\sqrt{3}} = 0.577.$$

$\therefore \quad \theta = \cos^{-1}(0.577) = 114°,$

For $\quad m_s = \frac{1}{2}$, we have

$$\cos\theta = -\frac{1}{\sqrt{3}} = -0.577.$$

$\therefore \quad \theta = \cos^{-1}(-0.557) = 125.3°.$

These angles are indicated in the diagram.

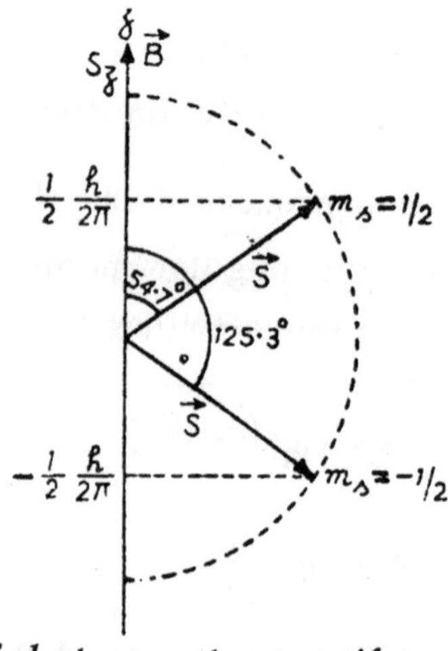

2. ***A beam of electrons enters a uniform magnetic field of 1.2 Tesla. Calculate the energy difference between electrons whose spins are parallel and anti-parallel to the field.***

Solution: An electron has an intrinsic (spin) angular momentum $\vec{S}$ and an intrinsic magnetic dipole moment $\vec{\mu}_s$, which are related by

$$\vec{\mu}_s = -\frac{e}{m}\vec{S},$$

where $$S = \sqrt{[s(s+1)]}\,\frac{h}{2\pi} \text{ and } s = \frac{1}{2}.$$

Let the magnetic field $\vec{B}$ be along the z-axis. The magnitude of the z-component of the magnetic moment is

$$\mu_x = \frac{e}{m} S_z$$

where S_z is the z-component of the spin angular momentum and is given by

$$S_z = m_s \frac{h}{2\pi},$$

where $m_s = \pm \frac{1}{2}$, depending upon whether S_z is parallel or anti-parallel to the z-axis. Thus

$$\mu_{sz} = \frac{e}{m} m_s \frac{h}{2\pi}$$

$$= \frac{e}{m}\left(\pm\frac{1}{2}\right)\frac{h}{2\pi} = \pm\frac{eh}{4\pi m}.$$

Now, the magnetic potential energy of a dipole of moment $\vec{\mu}_s$, in a magnetic field $\vec{B}$ is given by

$$V_m = -\vec{\mu}_s \cdot \vec{B}$$

$$= -\mu_{sz}\,\vec{B},$$

where μ_{sz} is the scalar magnitude of $\vec{\mu}_s$ in the direction of the magnetic field. Thus

$$V_m = \pm\frac{eh}{4\pi m} B.$$

The difference in energy of the electrons having spin parallel and anti-parallel to the field is

$$\Delta V_m = \frac{ehB}{4\pi m} - \left(-\frac{ehB}{4\pi m}\right) = \frac{ehB}{2\pi m}.$$

Substituting the known values of e, h and m; and the given values of B, we get

$$\Delta V_m = \frac{\left(1.6 \times 10^{-19}\ \text{coul}\right) \times \left(6.63 \times 10^{-34}\ \text{joule - sec}\right) \times \left(1.2\ \text{nt / amp} - \text{m}\right)}{2 \times 3.14 \times \left(9.1 \times 10^{-31}\ \text{kg}\right)}$$

$$= 2.23 \times 10^{-23}\ \text{joule.}$$

But $\quad 1\text{eV} = 1.6 \times 10^{-19}$ joule.

$$\therefore \quad \Delta V_m = \frac{2.23 \times 10^{-23}}{1.6 \times 10^{-19}} = 1.39 \times 10^{-4}\ \text{eV}.$$

3. ***(a) For a one-electron atom, write down the spectroscopic symbols for the possible energy levels of an electron with $l = 2$. (b) Which of these levels has the higher energy and why? (c) If the atom is placed in a weak magnetic field, into how many magnetic levels will each of the above levels split up? (d) Which one of these magnetic levels will have the highest energy and why ?***

Solution: (*a*) For the given election, we have

$$l = 2 \text{ and } s = \frac{1}{2}.$$

Therefore, the possible values of inner quantum number j are

$$j = l \pm s = 2 \pm \frac{1}{2} = \frac{5}{2}, \frac{3}{2}.$$

The multiplicity of the energy levels, defined by 2s + 1, is 2.

The energy levels for the election are designated by S, P, D,......according as $l = 0, 1, 2,$...... The multiplicity is indicated by a super-script and the value of j by a sub-script. Thus the possible energy levels for l = 2 electron in a single-electron atom will be written as

$$^2D_{5/2},\ ^2D_{3/2}.$$

(*b*) The level $^2D_{5/2}$ ($j = \frac{5}{2}$) corresponds to spin momentum parallel to orbital momentum while the level $^2D_{3/2}$ ($j = \frac{3}{2}$) corresponds to spin momentum anti-parallel to orbital momentum. Of these, the more stable level is one in which the magnetic moment of the electron spin, $\vec{\mu}_s$, lines up in the direction of the magnetic field $\vec{B}$ produced by the orbital motion of the electron.

Now, the field $\vec{B}$ due to orbital motion is always in the direction of angular momentum vector $\vec{L}$ (Fig.). Since the electron is negatively charged, the spin magnetic moment $\vec{\mu}_s$ is opposite to spin angular momentum $\vec{S}$. Thus $\vec{\mu}_s$ is in the direction of $\vec{B}$ for the level in which $\vec{S}$ is anti-parallel to $\vec{L}$ *i.e.* for the level corresponding to $j = l — s$ (Fig.). ($\vec{\mu}_s$ is opposite to $\vec{B}$ when $\vec{S}$ is parallel to $\vec{L}$ as in Fig.). Hence the level $^2D_{3/2}$ is more stable, *i.e.* of lower energy. The level $^2D_{5/2}$ is higher.

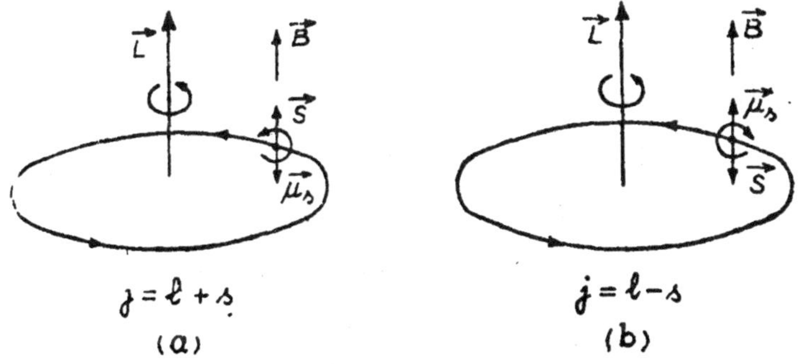

(*c*) When the atom is placed in a weak magnetic field, each energy level breaks up into $2j + 1$ magnetic levels corresponding to $mj = j, j—1, ...0, ...— j$. Thus the level $^2D_{5/2}$ breaks up into

6 magnetic levels $\left(m_j = \frac{5}{2}, \frac{3}{2}, \frac{1}{2}, -\frac{1}{2}, -\frac{3}{2}, -\frac{5}{2},\right)$ and the level $^2D_{3/2}$ breaks up into 4 magnetic levels $\left(m_j = \frac{3}{2}, \frac{1}{2}, -\frac{1}{2}, -\frac{3}{2},\right)$.

(*d*) The level corresponding to $m_j = \frac{5}{2}$ involves highest magnetic shift and hence lies highest.

4. *For an electron orbit with quantum number l = 2, state the possible values of the components of angular momentum along a specified direction.*

Solution: The component of total angular momentum along a specified direction (z-axis) is quantised, and takes values given by

$$J_z = m_j \frac{h}{2\pi}$$

where m_j is the magnetic quantum number corresponding to the inner quantum number j.

For the given electron, we have

$$l = 2 \text{ (d-electron)}$$

and, of course $s = \frac{1}{2}$.

The two possible values of j are

$$j = l \pm s$$

$$= 2 \pm \frac{1}{2} = 5/2 \text{ and } 3/2.$$

For $j = 5/2$ the possible values of mj are

$mj = 5/2, 3/2, 1/2, -1/2, -3/2, -5/2.$

For $j = 3/2$ the possible values of m_j are

$m_j = 3/2, 1/2, -1/2, -3/2.$

The values differ by integers. Thus, the possible values of the z-components of total angular momentum are

$$\pm\frac{5}{2}\left(\frac{h}{2\pi}\right), \pm\frac{3}{2}\left(\frac{h}{2\pi}\right), \pm\frac{1}{2}\left(\frac{h}{2\pi}\right).$$

5. ***Calculate the possible orientations of the total angular momentum vector $\vec{J}$ corresponding to j = 3/2 with respect to a magnetic field along the z-axis.***

Solution: The magnitude of the total angular momentum and its z-component are quantised according to the relations

$$J = \sqrt{[j(j+1)]}\,\frac{h}{2\pi}$$

and $$J_z = m_j\,\frac{h}{2\pi}$$

For j =3/2, we have

$$m_j = 3/2, 1/2, -1/2, -3/2.$$

The angle θ between $\vec{J}$ and the z-axis is determined by m_j and j, according as

$$\cos\theta = \frac{J_z}{J} = \frac{m_j}{\sqrt{[j(j+1)]}}$$

Now $$\sqrt{[j(j+1)]} = \sqrt{\left[\frac{3}{2}\left(\frac{3}{2}+1\right)\right]} = \frac{\sqrt{(15)}}{2}.$$

$$\therefore \qquad \cos\theta = \frac{2m_j}{\sqrt{(15)}}$$

For $$mj = 3/2, 1/2, -1/2, -3/2, \text{ we have}$$

$$\cos\theta = 0.775, 0.258, -0.258, -0.775$$

$$\therefore \qquad \theta = 39.2°, 75.0°,]114°, 140.8°.$$

The orientations of $\vec{J}$ with respect to z-axis are shown in Fig.

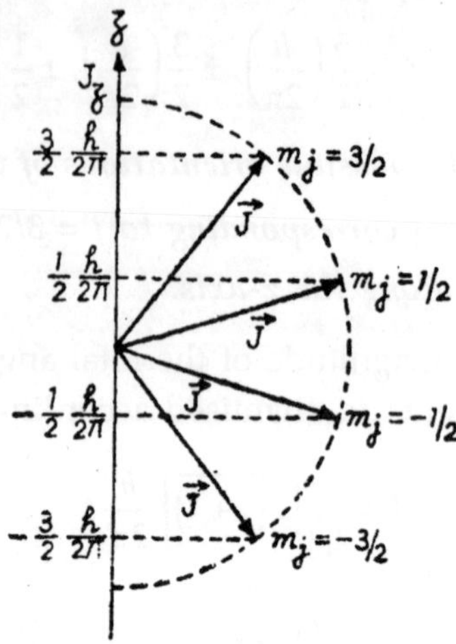

6. ***An element under spectroscopic examination is placed in a magnetic field of flux density 0.3 weber/meter². Calculate the Zeeman shift of a spectral line of wavelength 4500 Å.***

Solution: The Zeeman shift in frequency is given by

$$\Delta v = \frac{eB}{4\pi m}$$

where e and m are the charge and mass of electron. Since $v = \frac{c}{\lambda}$, we have

$$\Delta v = -\frac{c}{\lambda^2}\Delta\lambda$$

$$\therefore \qquad \Delta\lambda = \frac{\lambda^2}{c}\Delta v = \frac{eB\lambda^2}{4\pi mc}.$$

Here $B = 0.3$ weber/meter², $\lambda = 4500$ Å $= 45 \times 10^{-7}$ meter.

Thus

$$\Delta\lambda = \frac{\left(1.6 \times 10^{-19} \text{ coul}\right) \times \left(0.3 \text{ weber/meter}^2\right) \times \left(4.5 \times 10^{-7} \text{ m}\right)^2}{4 \times 3.14 \times \left(9.1 \times 10^{-31} \text{kg}\right)\left(3 \times 10^8 \text{meter/sec}\right)}$$

$$= 0.0283 \times 10^{-10} \text{ meter} = 0.0283 \text{ Å}.$$

7. The Zeeman components of a 500-nm spectral line are 0.0116 nm apart when the magnetic field is 1.00 T. Find the ratio e/m for the electron.

Solution: The Zeeman shift in wavelength is given by

$$\Delta\lambda = \frac{eB\lambda^2}{4\pi mc}.$$

From this, we have

$$\frac{e}{m} = \frac{4\pi c}{B}\left(\frac{\Delta\lambda}{\lambda^2}\right).$$

Here $\lambda = 500$ nm $= 5.00 \times 10^{-7}$ m, $\Delta\lambda = 0.0116$ nm $= 0.000116 \times 10^{-7}$ m, B =1.00, Tesla = 1.00 N/A-m and also c = 3.0×10^8 m/s.

$$\therefore \quad \frac{e}{m} = \frac{4 \times 3.14\left(3.0 \times 10^8 \text{ m/s}\right)\left(0.0001 \times 10^{-7} \text{ m}\right)}{\left(1.00 \text{ N/A} - \text{m}\right)\left(5.00 \times 10^{-7} \text{ m}\right)^2}$$

$$= 1.75 \times 10^{11} \text{ C/kg}.$$

11

Significance of Coherences

Various Coherences

A wave which appears to be a pure sine wave for an infinitely large period of time or in an infinitely extended space is said to be a perfectly coherent wave. In such a wave there is a definite relationship between the phase of the wave at a given time and at a certain time later, or at a given point and at a certain distance away. No actual light source, however, emits a perfectly coherent wave. Light waves which are pure sine waves only for a limited period of time or in a limited space are partially coherent waves.

There are two different criteria of coherence; the criteria of time and the criteria of space. This gives rise to temporal coherence and spatial coherence.

Role of Temporal Coherence

The oscillating electric field E of a perfectly coherent light wave would have a constant amplitude of vibration at any point, while its phase would vary linearly with time. As a function of time the field would appear as shown in Fig. It is an ideal sinusoidal function of time.

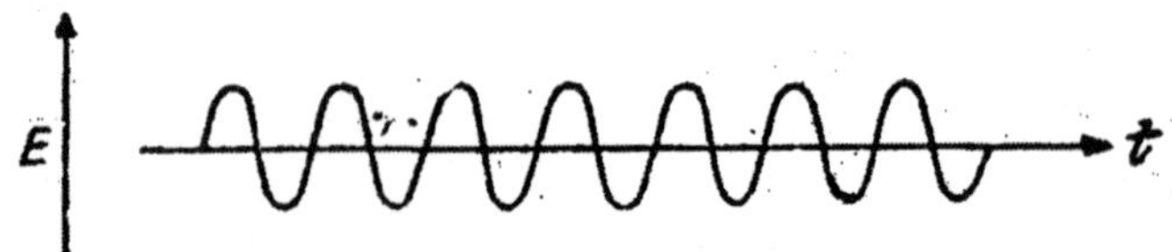

However, no light emitted by an actual source produces an ideal sinusoidal field for all values of time. This is because when an excited atom returns to the initial state, it emits light "pulse" of short duration such as of the order of 10^{-10} second for sodium atom. Thus, the field remains sinusoidal for time-intervals of the order of 10^{-10} second, after which the phase changes abruptly. Hence the field due to an actual light source will be as shown in Fig. The average time-interval for which the field remains sinusoidal (*i.e.* definite phase relationship exists) is known as "coherence time" or "temporal coherence" of the light beam, and is denoted by τ. The distance L for which the field is sinusoidal is given by.

$$L = \tau c,$$

where c is the speed of light. L is called the "coherence length" of the light beam.

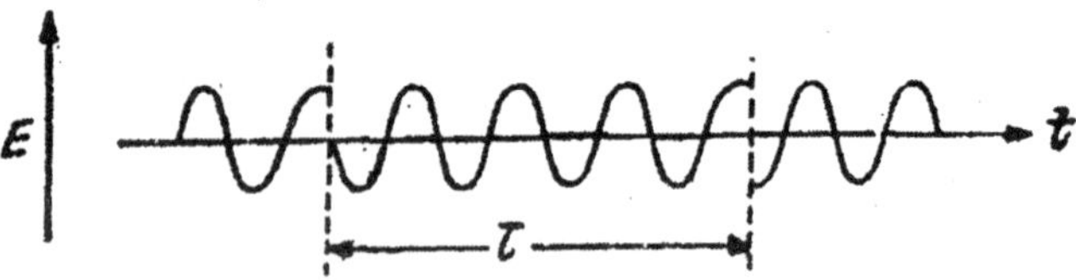

The coherence time (or the coherence length) can be measured by means of Michelson's interferometer (Fig.). A light beam from the source S falls on a semi silvered plate P at which it is partly reflected and partly transmitted. The reflected and transmitted beams, 1 and 2, are reflected back from mirrors M_1 and M_2 respectively and enter the telescope T in which interference fringes are observed. We know that the two beams can produce a stationary interference pattern only if there is a definite phase relationship between them.

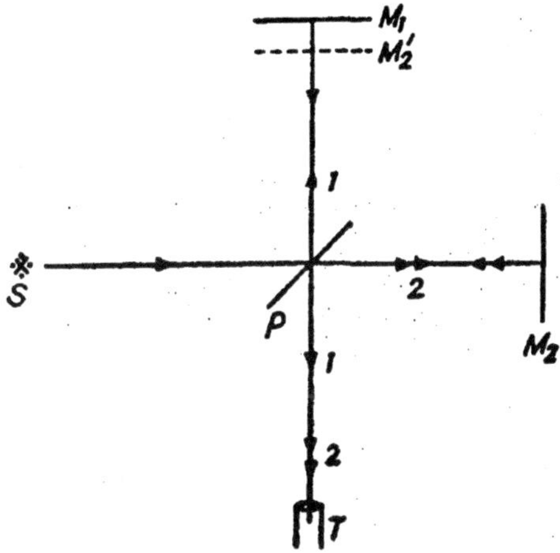

Let M_z' be the image of M_2 formed by the plate P. The arrangement is then equivalent to an air-film enclosed between two reflecting surfaces M_1 and M_2'. If d is the separation between M_1 and M_2', then $2d$ will be the path difference between the interfering beams. Now, if

$$2d << L,$$

then there will be a definite phase relationship between the two beams and interference fringes will be observed. If, on the other hand,

$$2d >> L,$$

there will be no definite phase relationship and the fringes will not be observed. Therefore, starting with equal path-lengths, as the distance d is increased (by moving one mirror), the fringes become gradually poorer in contrast and finally disappear. The path difference at disappearance gives an estimation of coherence length.

For the sodium yellow light the coherence length is about 3 cm, so that the coherence time is

$$\tau = \frac{L}{c} \approx \frac{3\text{cm}}{3 \times 10^{10}\ \text{cm/sec}} \approx 10^{-10}\ \text{sec.}$$

For the Krypton orange light the coherence length is about 30 cm while for laser light it can be few kilometers.

Role of Spatial Coherence

The spatial coherence is the phase relationship between the radiation fields at different points in space. Let us consider light waves emitting from a source S (Fig.). Let *A* and *B* be two points lying on a line joining them with S. The phase relationship between *A* and *B* depends on the distance *AB* and on the temporal coherence of the beam. If $AB << L$ (coherence length), there will be a definite phase relationship between *A* and *B*, i.e. there will be high coherence between *A* and *B*. On the other hand, if $AB >> L$, there will be no coherence between *A* and *B*.

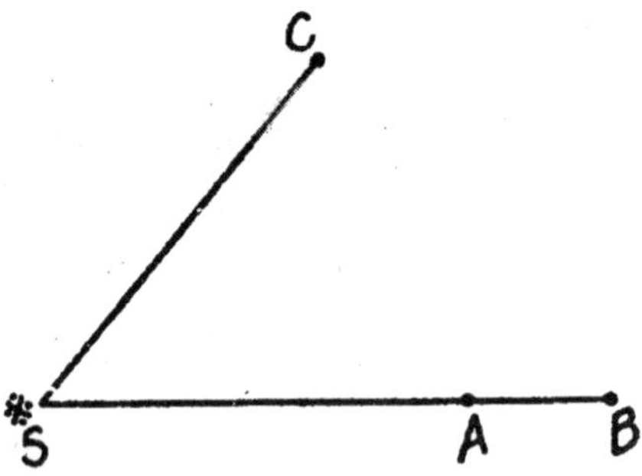

Let us now consider points *A* and *C* which are equidistant from S. If the source *S* is a true *point* source, then the waves shall reach *A* and C in exactly the same phase, *i.e.* the two points will have perfect (spatial) coherence. If, however, the source S is extended, the points *A* and *C* will no longer remain in coherence. This may be emonstrated by Young's double-slit experiment illustrated in Fig. Light emitting from a narrow slit *S* falls on two slits S_1 and S_2 placed symmetrically with respect to *S*. The beams emerging from S_1 and S_2, having been derived from the same original beam, maintains a constant phase difference at all points on the screen. Hence a stationary interference pattern is observed on the screen.

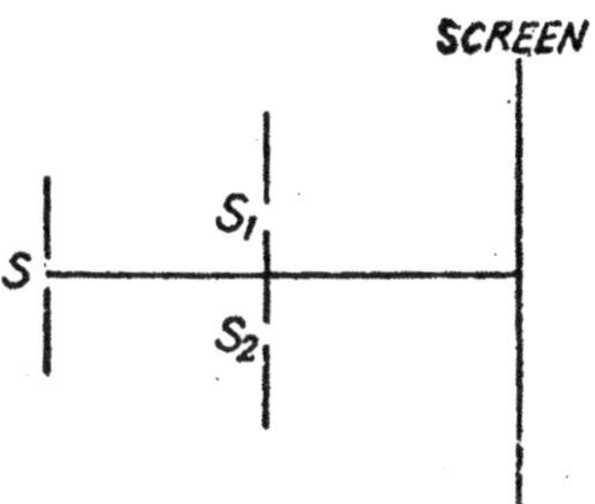

If, however, the width of the slit *S* is gradually increased, the pattern becomes poorer and poorer in contrast and finally disappears. This means that as the size of the source is increased, the situation of spatial coherence on the screen changes into a situation of incoherence. This happens because when S is wide, S_1 and S_2 receive waves; from different independent parts of *S* and hence do not remain coherent with respect to each other.

We may derive a relationship between the spatial coherence and the size of source. An extended source is made up of a largo number of point-sources. Let us first consider the case when the Young's double slit is illuminated by two independent point-sources *S* and *S'* at a distance *l* apart (Fig.). We shall find minimum value of *l* at which pattern on the screen would disappear. The waves from S which reach the point *O* on the screen *via* S_1 and S_2. have zero path difference. Hence there is a bright fringe at *O* due to S. Now, the waves from *S'* reaching the point *O via* S_1 and S_2 have a path difference $S'S_2$ -$S'S_1$ = KS_2. Clearly, we shall obtain a dark fringe at *O* due to *S'* when

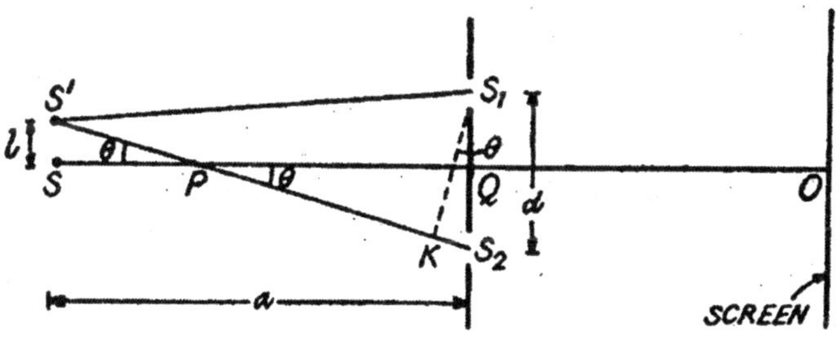

$$KS_2 = \frac{\lambda}{2}, \qquad \text{... } (i)$$

where λ is the wavelength of light. When this is the case, the maxima of the interference pattern due to S will fall on the minima due to S' so that the fringes would disappear. Now, from the figure,

$$KS_2 \simeq \theta d,$$

where d is the separation between S_1 and S_2. Let a be the distance between S and Q. Now, again from the figure,

$$\theta \simeq \frac{QS_2}{PQ} \simeq \frac{SS'}{SP}.$$

$$\therefore \qquad a = SP + PQ \simeq \frac{SS'}{\theta} + \frac{QS_2}{\theta} = \frac{l}{\theta} + \frac{d/2}{\theta}$$

or $$\theta \simeq \left(l + \frac{d}{2}\right) / \text{a}.$$

$$\therefore \qquad KS_2 \sim \theta d \simeq \left(l + \frac{d}{2}\right) d / a.$$

Assuming that $l >> \frac{d}{2}$; we can write.

Thus in view of eq. (*i*), the interference pattern would disappear if,

$$\frac{ld}{a} \simeq \frac{\lambda}{2}$$

or $$\lambda \sim \frac{\lambda a}{2d} \qquad \text{... } (ii)$$

From this it is clear that if we illuminate the double-slit with an extended source whose linear dimension exceeds $\lambda a/2d$, then no interference pattern will appear on the screen.

If the extended source S' subtends an angle α at Q' then

$$\propto \simeq \frac{l}{a}$$

Eq. (*ii*) can be rewritten as

$$d \simeq \frac{\lambda a}{2l} \simeq \frac{\lambda}{2\alpha}.$$

$\frac{\lambda}{\alpha}$ is termed as "lateral spatial coherence width". We conclude that to obtain a good interference pattern with Young's double-slit, the separation between the slits (S_1 and S_2) should be kept much less than the coherence width.

Purity of a Spectral Line: Every spectral line has a finite width which means that it corresponds to a continuous distribution of wavelengths in some narrow interval between λ and $\lambda + \Delta\lambda$. This is obvious from the fact that for every spectral line the interference pattern in the Michelson interferometer experiment eventually disappears when the path difference between the interfering beams is gradually increased. We also know that the pattern disappears when the path difference exceeds the coherence length. Thus the concept of temporal coherence is directly related to the width (or purity) of the spectral line.

Let us adopt the criterion that when the path difference between the interfering beams becomes equal to the coherence length L, the rings due to two closely spaced wavelengths λ_1 and λ_2 are completely out of step at the centre (*i.e.* a bright ring of λ_1 coincides with a dark ring of λ_2). Then we can write

$$L = n\lambda_1 = (n + \frac{1}{2})\,\lambda_2.$$

Eliminating n, we get

$$L = \left(\frac{L}{\lambda_1} + \frac{1}{2}\right)\lambda_2$$

or $$\frac{L}{\lambda_2} - \frac{L}{\lambda_1} = \frac{1}{2}$$

or $$L\,\frac{\lambda_1 - \lambda_2}{\lambda_1\lambda_2} = \frac{1}{2}$$

or $$L = \frac{\lambda_1\lambda_2}{2(\lambda_1 - \lambda_2)}$$

If instead of two discrete wavelengths λ_1 and λ_2, the beam consists of all wavelengths lying between λ_1 and λ_2, then the pattern would disappear if

$$L = \frac{\lambda_1\lambda_2}{\lambda_1 - \lambda_2}$$

or $$L \simeq \frac{\lambda_2}{\Delta\lambda}$$

or $$\Delta\lambda \simeq \frac{\lambda^2}{L}$$

Thus if the fringes become indistinct when the path difference exceeds *L*, we can conclude that the spectral line (of mean wavelength λ) has a wavelength-spread (width) given by $$\Delta\lambda \simeq \frac{\lambda^2}{L}.$$

For the cadmium red line (λ = 6438 Å) *L* is as large as about 30 cm. This corresponds to a wavelength-spread given by

$$\Delta\lambda \simeq \frac{\lambda^2}{L} \simeq \frac{(6438\text{ Å})^2}{30 \times 10^8\text{ Å}} \simeq 0.01\text{ Å}.$$

Further, since $v = \frac{c}{\lambda}$, the frequency-spread Dv of a line would be

$$\Delta v = \frac{c}{\lambda^2}\Delta\lambda \simeq \frac{c}{L}$$

We know that $\tau \simeq \frac{L}{c}$ where τ is coherence time.

$$\therefore \qquad \Delta v \sim \frac{1}{\tau}.$$

Thus the frequency-spread of a spectral line is of the order of the inverse of the coherence time. It means that a perfectly sharp monochromatic line ($\Delta v - 0$) would correspond to an infinite interval of time ($\tau = \infty$).

12

Statistical Mechanics

Phase Space

In classical mechanics, the dynamical state of a particle at a particular instant is completely specified if its three position coordinates x, y, z and three velocity, or preferably momentum components p_x, p_y, p_z, at that instant are known. This conception is generalised by imagining a six-dimensional space in which a point has six coordinates (x, y, z, p_n, p_y, p_z). Such a space is called 'phase space'. The instantaneous state (position and momentum) of a particle is represented by a point in the phase space.

For a system containing a large number of particles, there would be a large number of representative points in the phase space corresponding to the instantaneous distribution of particles. Thus the state of a system of particles corresponds to a certain distribution of points in phase space.

Let us divide the phase space into tiny six-dimensional cells whose sides are $dx, dy, dz, dp_x, dp_y, dp_z$. The volume of each cell is

$$d\tau = dx\ dy\ dz\ dp_x\ dp_v\ dp_z.$$

$d\tau$ is termed as 'an element of volume in the phase space'. $dx\ dy\ dz$ is an element of volume in coordinate space and $dp_x\ dp_y\ dp_z$ is an element of volume in momentum space.

Now, according to the uncertainty principle, we have

$$dx\ dp_x \geq \frac{h}{4\pi}\ ;\ dy\ dp_y \geq \frac{h}{4\pi}\ ;\ dz\ dp_z \geq \frac{h}{4\pi}$$

so that
$$d\tau \geq \left(\frac{h}{4\pi}\right)^3.$$

A more detailed analysis shows that

$$d\tau = h^3.$$

Thus, due to uncertainty principle, a "point" in phase space is actually a cell of volume h^3. A point (x, y, z, p_x, p_y, P_z) in the phase space representing the position and momentum of a particle specifies that the particle lies somewhere in a cell of volume A^8 centered at the point.

The notion of phase space is important in describing the behaviour of a system of particles. The equilibrium state of the system corresponds to the most probable distribution of particles in the phase space.

Statistical Mechanics: Techniques

Every solid, liquid or gas is a system (or an assembly) consisting of" an enormous number of microscopic particles. Likewise, radiation is an assembly of photons. Obviously, the actual motions or interactions of individual particles cannot be investigated. However, the macroscopic properties of such systems can be explained in terms of the most probable behaviour of the individuals. By employing statistical methods we can determine how the individuals of a given assembly are distributed among different possible states and what is their most probable behaviour.

Let us consider statistically how a fixed amount of energy is distributed among the various individuals of an assembly of *identical* particles. There are three kinds of identical particles:

(i) Identical particles of any spin which are so much separated in the assembly that they *can be distinguished* from one another. The molecules of a gas are particles of this kind. The Maxwell-Boltzmann distribution holds for these particles.

(ii) Identical particles of 0 or integral spin which *cannot be distinguished* from one another. These are called Bose particles (or Bosons) and do not obey Pauli's exclusion principle. Photons, phonons and α-particles are of this kind. The Bose Einstein distribution holds for them.

(iii) Identical particles of odd half-integral spin which *can not be distinguished* from one another. These are called Fermi particles (or Fermions) and do obey Pauli's exclusion principle. Electrons, protons and neutrons are particles of this kind. The Fermi-Dirac distribution holds for them.

Distribution Law

The energy distribution law for particles of kind (*i*) can be derived by methods of classical statistics as well as of quantum statistics, but that for particles of kinds (*ii*) and (*iii*) can be derived by methods of quantum statistics only.

Maxwell-Boltzmann (Classical) Distribution: Let us consider a system composed of a very large number N of *distinguishable* identical particles. Suppose the energies of the particles are limited to the values ε_1, ε_2,......, ε_r which represent either discrete energy states or average energies within a sequence of energy intervals. Let us divide the whole volume of the phase space into r cells and distribute the N molecules among these cells. Let us consider a distribution of the molecules with respect to their energy such that n_1, molecules (each having energy ε_1) are in cell 1, n_g molecules (each having energy ε_2) are in cell 2, and so on. The probability W for the distribution is given by

$$W = \frac{N!}{n_1 ! n_2 ! \ldots\ldots n_r !} (g_1)^{n_1} (g_2)^{n_2} \ldots \ldots (g_r)^{n^r}$$

$$= \frac{N!}{\Pi n_i !} \Pi (g_t)^{n_{i,}}$$ [just as Σ stands for sum Π stands for product]

where g_i is the a *priori probability* for a particle to have the energy ε_1.

There are two limitations. The total number of particles is fixed at N. This means that

$$\sum_{i-1}^{r} nt = n_1 + n_2 + \ldots \ldots + nr = N.$$

At a constant temperature, the total energy content is fixed at E (say). Thus

$$\sum_{i-1}^{r} nt\ \varepsilon i = n_1 \varepsilon_1 + n_2 \varepsilon_2 + \ldots \ldots + n_r \varepsilon_r = E.$$

The next step is to determine the most probable distribution of the particles which would correspond to the state of thermodynamic equilibrium. This distribution would yield maximum value of W. Thus, for the equilibrium state, we have

$$\delta W = 0$$

subject to the limitations

$$\Sigma \delta n_i = 0$$

and $$\Sigma \varepsilon_i \delta n_i = 0$$

This leads to the result

$$n_i = \frac{gi}{e^{\alpha} e^{\varepsilon i/kT}},$$

where e^{α} is a constant, k is Boltzmann's constant and T is Kelvin temperature. This result is known as Maxwell-Boltzmann distribution law.

Bose-Einstein (Quantum) Distribution: The Maxwell-Boltzmann distribution governs identical particles which can be distinguished from one another in some way (as molecules in a gas) and as such they can be given names or numbers. On the other hand, Bose-Einstein distribution governs identical particles which can never be distinguished from one another (as radiation photons in an enclosure). As such there is no way to number them or give them names.

The basic assumption of the B-E distribution is that any number of particles can be in any quantum state and that all quantum states are equally probable. Suppose each quantum state corresponds to an elementary cell in the phase space. The number of different ways in which n_i *indistinguishable* particles can be distributed in g_i phase-cells is

$$\frac{(n_i + gi - 1)^!}{ni!(gi - 1)^!}$$

When the number of cells is sufficiently large, this is same as

$$\frac{(n_i + gi)^!}{ni!\,gi^{\,!}}$$

The probability W of the entire distribution of N particles is the *product of* the numbers of different arrangements of particles among the states having each energy. Thus

$$W = \Pi\,\frac{(n_i{}' + gi)^!}{ni!\,gi^{\,!}}$$

For the equilibrium (most probable) condition, we must have

$$\delta W = 0$$

subject to the limitations

$$\Sigma\, \delta n_i = 0$$

and $$\Sigma\, \varepsilon i\, \delta n_i = 0.$$

This leads to the result

$$ni = \frac{gi}{e^{\alpha}\, e^{\epsilon i/kT} - 1}$$

where e^{α} is a constant, k is Boltzmann's constant and T is Kelvin temperature. This is Bose-Einstein distribution law.

Fermi-Dirac (Quantum) Distribution: Fermi-Dirac distribution applies to *indistinguishable* particles, like electrons, which are governed by Pauli's exclusion principle. The general method of deriving the Fermi-Dirac distribution law is similar to that for the Bose-Einstein statistics except that now each phase-cell (that is, each quantum state) cannot be occupied by more than one particle. This leads to the following distribution function:

$$ni = \frac{gi}{e^{\alpha}\, e^{\epsilon i/kT} + 1}.$$

Again, n_i is the number of particles whose energy is ε_i, and g_i is the number of states that have the same energy εi.

Comparison of the Three Distribution Laws: For this purpose, let us define a quantity

$$f(\varepsilon_i) = \frac{ni}{gl}.$$

$f(\varepsilon_i)$ is called the 'occupation index' of a state of energy ε_i. It represents the average number of particles in each of the states of that energy. Thus

$$f_{MB}\,(\varepsilon i) = \frac{1}{e^{\alpha}\, e^{\epsilon i/kT}}$$

$$f_{ME}\,(\varepsilon i) = \frac{1}{e^{\alpha}\, e^{\epsilon i/kT} - 1}$$

and $$f_{FD}\,(\varepsilon i) = \frac{1}{e^{\alpha}\, e^{\epsilon i/kT} + 1}$$

In the following diagrams we plot each occupation index $f(\varepsilon_i)$ against energy ε_i for two different values of T and α.

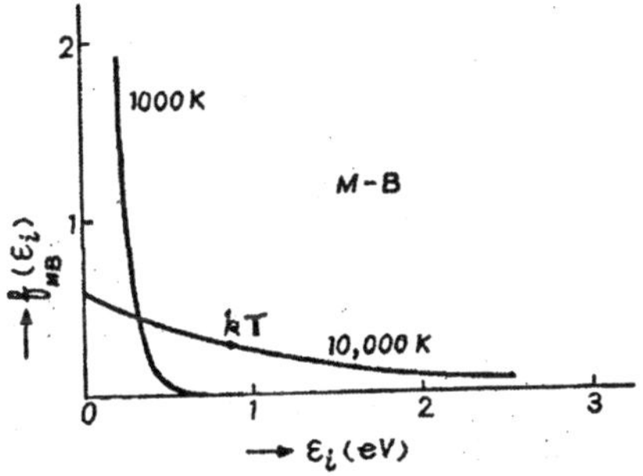

In the M-B distribution (Fig.) the occupation index $f(\varepsilon_i)$ falls purely *{exponentially* with increase in energy ε_i. In fact, it falls by a factor of $\frac{1}{e}$ for each increase of kT in the energy $\in_i$.

The B-E occupation index against energy is plotted in the figure below for temperatures of 1000 K and 10,000 K, in each case for $\alpha = 0$ (corresponding to a "gas" of photons). For $\in_i >> kT$, the B-E distribution approaches the exponential form characteristic of the M-B distribution. In this region the average number of particles per quantum state is much less than one. However, for $\in_i << kT$, the —1 term in the denominator causes the occupation index much greater compared to that of the M-B distribution. This means that for energies small compared to kT, the number of particles per quantum state is greater for the B-E distribution than for the M-B distribution.

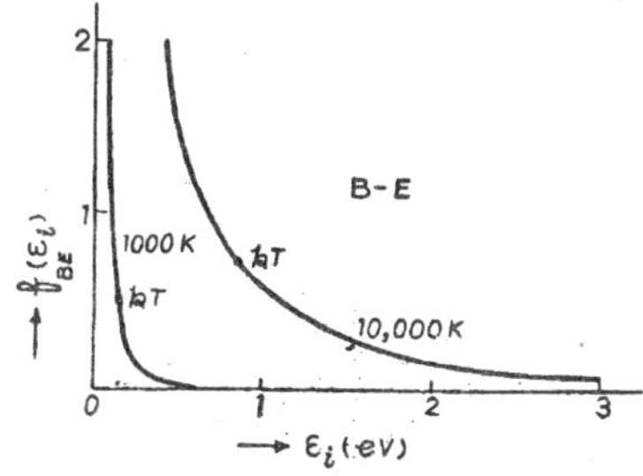

The F-D distribution is plotted in Fig. below for three different values of T and α. In this distribution the occupation index never goes above 1. This signifies that we cannot have more than one particle per quantum state as required by Pauli's exclusion principle which applies in this case. Further, in this distribution, the parameter α is strongly dependent on temperature T, and we write:

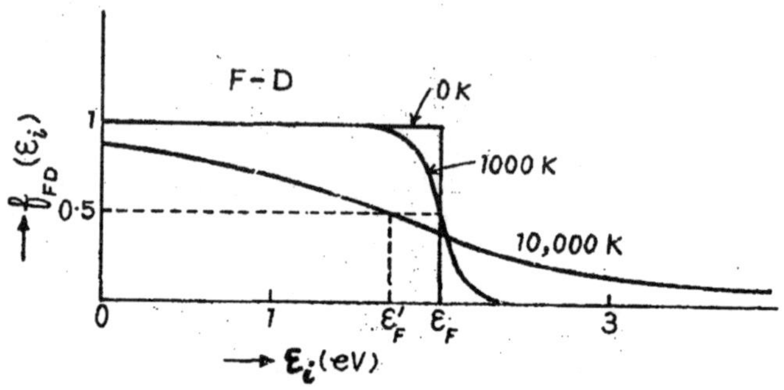

$$\alpha = -\frac{\in F}{kT}$$

so that the F-D occupation index becomes

$$f_{FD}(\in i) = \frac{1}{e^{(\in i - \in F)/kT} + 1},$$

where $\in_F$ is called the 'Fermi-energy'.

Let us consider the situation at the absolute zero of temperature. At $T = 0$, $(\in_i - \in_F)/kT = -\infty$ (for $\in_i < \in_F$) and $+\infty$ (for $\in_i > \in_F$), so that

$$\text{for } \in_i < \in_F, f_{FD}(\in_i) = \frac{1}{e^{-\infty} + 1} = 1, \quad (\because e^{-\infty} = 0)$$

$$\text{and for } \in_i > \in_F, f_{FD}(\in_i) = \frac{1}{e^{+\infty} + 1} = 0. \quad (\because e^{+\infty} = \infty)$$

Thus, at $T = 0$, all energy states from $\in_i = 0$ to $\in_i = e_F$ are occupied because $f_{FD}(\in_i) = 1$; while all above $\in_F$ are vacant.

As the temperature rises, some of the states just below $\in_F$ become vacant while some just above $\in_F$ are occupied. The higher the temperature, the more is the spreading in $f_{FD}(\in_i)$.

At $\in_i = \in_F$, we have

$f_{FD}(\in_i) \dfrac{1}{e^0 + 1} = ½$, at all temperatures, that is, the average number of particles per quantum state is exactly ½. In other words, *the probability of finding an electron with energy equal to the Fermi energy in a metal is ½ at any temperature.*

Bibliography

Alam, T.: *Physical Principles of Flow in Unsaturated Porous Media*, Oxford, New York, 1994.

Arun, Kumar: *Problems in Physics*, Atlantic Publishers, New Delhi, 2003.

Atherton, J.C.: *Thermal Physics*, Butterworths, London, 1987.

Bowling, D.J.F.: *Basic Physics*, Chapmann and Hall, London, 1976.

Caro, D. E.: *Modern Physics: an Introduction to Atomic and Nuclear Physics*, Edward Arnold, London, 1985.

Cartwright, H.R.: *Elements of Thermal and Statistical Physics*, Oxford University Press, London, 1966.

Crafts, A.S.: *Phloem Transport in Plants*, W.H. Freeman, San Francisco, 1973.

Daniel, W.: *Introduction to Physics*, MacMillan Press, Melbourne, 1969.

Divan, T.: *Thermal Analysis of Materials*, W.A. Benjamin, California, 1975.

Donald, J.S.: *Advanced Guide to Modern Physics*, Pelham Books, London, 1976.

Epstein, E.: *Statistical Physics*, John Wiley and Sons, New York, 1972.

Esau, K.: *Physical Geology*, John Wiley and Sons, New York, 1965.

Feynman and Others: *The Feymman Lectures on Physics,* Add. Wesley, London, 1989.

Feynman, R. P. and Others: *Elementary Particles and the Laws of Physics,* Cambridge, New York, 2002.

Francia, G.: *The Investigation of the Physical World,* Cambridge, New York, 1981.

Ftzler M.E.: *Plant Lectins: Molecular and Biological Aspects,* Academic Press, New York, 1985.

George, M.: *Elements of Physical Hydrology,* Johns Hopkins, New York, 1998.

Green, J.R.: *Modern Physics,* MacMillan, London, 1911.

Halevy, A.H.: *Handbook of Physics,* C.R.C. Press, Boca Raton, 1985.

Hart, J.W. and Sabnis, D.D.: *Encyclopaedia of Stastical Physics,* Springer Verlag, Berlin, 1982.

Hopkins, C.R.: *Structure and Function of Cells,* W.B. Saunders Co., London, 1978.

Jorgensen, T. P.: *The Physics of Golf,* Springer, New York, 1999.

Karl, F.: *Basic Physics,* John Wiley, London, 1996.

Kinet, J.M.: *Thermal and Statistical Physics,* C.R.C. Press, Boca Raton, 1981.

Kirkby, E.A.: *Mathematical Physics,* The University of Leeds, Leeds, 1970.

Lain, G.: *Vibrations and Waves in Physics,* Cambridge, New York, 1998.

Landshoff, P.: *Essential Quantum Physics,* Cambridge, New York, 2001.

Lipson, S. G.: *Optical Physics,* Cambridge, New York, 1995.

Lockett, Keith: *Physics in the Real World,* Cambridge, New York, 1995.

Lui, Lam: *Introduction to Nonlinear Physics,* Springer, New York, 1997.

McKee, H.S.: *University Physics,* Oxford University Press, London, 1962.

Meidner, H. and Sheriff, D.W.: *Water and Plants,* Blackie and Sons, Glasgow, 1976.

Mengel, K. and Kirkby, E.A.: *Thermal and Statistical Physics,* Praeger Publishers, New York, 1974.

Mercer, E.I.: *Recent Advances in the Chemistry and Biochemistry of Plant Lipids,* Academic Press, New York, 1975.

Minoru, Fujimoto: *The Physics of Structural Phase Transitions,* Springer, London, 2000.

Nabapro, F.R.: *The Physics of Creep,* T. & F. Publications, New York, 1995.

Negele, J. W. and Others: *Advances in Nuclear Physics,* Plenum, London, 1994.

Palladin, V.I.: *Physical Metallurgy,* Arihant Publishers, Jaipur, 1988.

Peter, Haasen: *Optical Physics,* Cambridge, New York, 1996.

Plischke, M. and Others: *Equilibrium Statistical Physics,* World Scientific, New York, 1994.

Pruce, D.: *Elements of Stastical and Thermodynamics Physics,* McGraw Hill, London, 1975.

Rabinowitch, E.: *Thermodynamics and Statistical Physics,* John Wiley, New York, 1969.

Rawn, J.D.: *Encyclopaedia of Modern Physics,* Harper and Row, New York, 1983.

Ross, C.W. and Salisbury, F.B.: *Thermal Physics,* Woodworth Publishing Co., California, 1967.

Rossi, Paolo: *The Birth of Modern Science,* Blackwell, New Jersey, 2001.

Ruhland, W.: *Encyclopaedia of Thermodynamics Physics,* Springer-Verlag, Berlin, 1961.

Sagdeev, R.Z.: *Nonlinear Space Plasma Physics,* AIP, New Jersey, 1993.

Schein, L. B.: *Electro-Photography and Development Physics,* Springer, London, 1992.

Slatyer, R.O.: *The Physics of Structural Phase Transitions,* Academic Press, New York, 1967.

Srivastava, H.S.: *Elements of Physics,* Rastogi Publications, Meerut, 1990.

Thimann, K.V.: *Physics and the Real World,* CRC Press, Boca Raton, Florida, 1980.

Tolbert, N.E.: *Microbodies—Peroxisomes and Glyoxysomes,* Oxford University Press, London, 1971.

Tommy, L.: *Modern Physics in Twenty first Century,* Springer Verlag, Berlin, 1975.

Tucker, G.A.: *Physics and Chance,* Oxford, New York, 1999.

Verma, W.A.: *Problems in Physics,* Academic Press, New York, 1970.

Verner, J.E.: *Thermal Physics,* Academic Press, New York, 1985.

William, R.: *Techniques for Nuclear and Particle Physics Experiments: A How to Approach,* Springer, London, 1994.

Young, H. D.: *University Physics,* Add. Wesley, London, 1996.

Zeevart, J.A.: *Thermal Plasma and New Materials Technology,* McGraw Hill, London, 1976.

Zenic, B.: *The Geometry and Physics of Knots,* Cambridge, New York, 1991.

Index

F

G

O

P

Q

R

S

T

U

V

W

X

Z